D0536109

MEDIA, RISK AND SCIENCE

WITHDRAWN
UTSA LIBRARIES

ISSUES in CULTURAL and MEDIA STUDIES

Series editor: Stuart Allan

Published titles

News Culture
Stuart Allan

Media, Risk and Science
Stuart Allan

Television, Globalization and Cultural Identities
Chris Barker

Cultures of Popular Music
Andy Bennett

Masculinities and Culture
John Beynon

Cinema and Cultural Modernity
Gill Branston

Ethnic Minorities and the Media
Edited by Simon Cottle

Modernity and Postmodern Culture
Jim McGuigan

Sport, Culture and the Media
David Rowe

Compassion, Morality and the Media
Keith Tester

MEDIA, RISK AND SCIENCE

Stuart Allan

OPEN UNIVERSITY PRESS
Buckingham · Philadelphia

Open University Press
Celtic Court
22 Ballmoor
Buckingham
MK18 1XW

email: enquiries@openup.co.uk
world wide web: www.openup.co.uk

and
325 Chestnut Street
Philadelphia, PA 19106, USA

First Published 2002

Copyright © Stuart Allan 2002

All rights reserved. Except for the quotation of short passages for the purpose of criticism and review, no part of this publication may be reproduced, stored in a retrieval system, or transmitted, in any form or by any means, electronic, mechanical, photocopying, recording or otherwise, without the prior written permission of the publisher or a licence from the Copyright Licensing Agency Limited. Details of such licences (for reprographic reproduction) may be obtained from the Copyright Licensing Agency Ltd of 90 Tottenham Court Road, London, W1P 0LP.

A catalogue record of this book is available from the British Library

ISBN 0 335 20662 X (pb) 0 335 20663 8 (hb)

Library of Congress Cataloging-in-Publication Data
Allan, Stuart, 1962–
Media, risk, and science/Stuart Allan.
p. cm. – (Issues in cultural and media studies)
Includes bibliographical references and index.
ISBN 0-335-20663-8 – ISBN 0-335-20662-X (pbk.)
1. Scence news. 2. Science in mass media. I. Title. II. Series.

Q225 A45 2002
500–dc21

2002023858

Library
University of Texas
at San Antonio

Typeset by Type Study, Scarborough
Printed in Great Britain by Biddles Limited, Guildford and Kings Lynn

CONTENTS

SERIES EDITOR'S FOREWORD

The Issues in Cultural and Media Studies series aims to facilitate a diverse range of critical investigations into pressing questions considered to be central to current thinking and research. In light of the remarkable speed at which the conceptual agendas of cultural and media studies are changing, the authors are committed to contributing to what is an ongoing process of re-evaluation and critique. Each of the books is intended to provide a lively, innovative and comprehensive introduction to a specific topical issue from a fresh perspective. The reader is offered a thorough grounding in the most salient debates indicative of the book's subject, as well as important insights into how new modes of enquiry may be established for future explorations. Taken as a whole, then, the series is designed to cover the core components of cultural and media studies courses in an imaginatively distinctive and engaging manner.

Stuart Allan

ACKNOWLEDGEMENTS

Many kind people helped to make this book much more engaging to write than it would have been without their shared support, enthusiasm and insights. I would like to thank the scientists, journalists, activists and policy-makers who agreed to be interviewed, and for tolerating my awkward questions. I'm grateful to Cynthia Carter, Felicity Mellor and Peter Broks for commenting on an early draft. For inviting me to their respective universities to help me test out some ideas in public talks, I thank Alison Anderson, David Berry, Donald Matheson, Martin Montgomery, Alvaro Pina, Karen Ross, Imogen Tyler and Chris Weedon. A further debt is owed to my fellow members of the Risk, Media and Communication Research Group, an inter-university initiative based at Cardiff University, for invigorating discussions. For also sharing their thoughts along the way, I'm grateful to Ulrich Beck, Frank Burnet, Simon Cottle, John Durant, Nils Lindahl Elliot, Peter Glasner, Olga Guedes, Iain Hamilton Grant, Dick Heller, Jim McGuigan, Adam Nieman, Lesley Patterson, Roberta Pearson, Chris Philippidis, Helen Lawton Smith, Joost van Loon, Ian Welsh, members of Harvard University's Center for Risk Analysis, and fellow participants at various conferences over the years. A special word of thanks to Barbara Adam, Martin Lister, Maureen McNeil and John Tulloch for their support. My colleagues and students in the School of Cultural Studies at the University of the West of England, Bristol continue to inspire me with their passion for pushing cultural and media studies in new directions. I'm similarly grateful to my faculty's Research Committee for a term's research leave, and to the Arts and Humanities Research Board (AHRB) for funding a further term's worth of research and writing. Justin Vaughan and colleagues at Open University Press have been brilliant, and I'm indebted to Christine Firth for her astute copyediting. And finally, as always, my heartfelt gratitude to my partner Cindy and son Geoff for everyday acts of kindness, large and small.

Stuart Allan

INTRODUCTION: MEDIA, RISK AND SCIENCE

> Most voters are frankly mere info-peasants, scientific illiterates, vacant idiots at the mercy of glossy corporate-science propaganda and newspaper hysterias. They are told a 'government scientist' is an authority, whether he's spent his life on earthworms or planets. They don't ask about peer-group review. They don't even have a clear notion of scientific proof, or the simple big discoveries that lead to the front-page stories that shock them.
>
> (Andrew Marr, journalist)

The world of science, judging from some media portrayals, is a world of white-coated boffins peering through microscopes, laboratory benches with bubbling flasks set above flickering Bunsen burners, and racks of test tubes and petri dishes emitting strange aromas. It is an insular world, cut off from the real world outside the laboratory window. The media tell us that this world of science is mind-numbingly boring and mundane in the repetition of its daily routines. Except, that is, for those rare moments when with a terrifyingly abrupt flash of insight (in the time it takes to, say, split an atom) the very future of humankind might suddenly appear to be hanging in the balance. If in the wrong hands science can be used to evil ends, in the right hands it is proclaimed to be our salvation. Scientists themselves tend to be represented as being decent, high-minded citizens tirelessly committed to the eternal pursuit of truth on behalf of their society. This endeavour, it follows, is a cornerstone of modern democracy, helping to make the world an organized, ordered system. Still, we are warned, there are exceptions. Lurking among their ranks are those intent on exploiting scientific knowledge for ominous purposes. These scientists, having been corrupted by greed or driven mad by a lust for power, are dangerously out of control.

The world of the media, at least according to statements sometimes made by scientists, is a superficial world driven by a frenzied obsession with entertainment over information, and with it style over substance. This is a world of smoke and mirrors, where nothing is as it seems, and where talk of ratings, target audiences and financial profits all but silences the voices of scientific truth. Journalists struggling to report on a scientific development, no matter how well intentioned they may be, will more often than not succumb to the forces of sensationalism to make their news account attract the public's wandering eye. If it bleeds, it leads. By the same logic, scientific facts must not be allowed to get in the way of a good story. Across the media, in-depth discussions and debates about scientific inquiries are up against, and losing out to, talking dinosaurs befriending humans, magicians happily breaking the laws of physics, mystics foretelling lottery results in their crystal balls, and horoscopes revealing people's fate and fortune as dictated by the stars. For some scientists, this recurrent misrepresentation of the scientific world by certain members of the media is more than just scandalous, it is contrary to the fundamental values and interests of democracy itself. The media's failure to give science the respect it deserves, they warn, will have dire consequences for the future.

It goes without saying, of course, that these two 'worlds' are being deliberately sketched here with broad and colourful brushstrokes. My intention in doing so is to highlight how the often subtle boundaries demarcating what counts as 'science' in a modern society need to be situated in relation to the kinds of images one typically encounters in the media. More to the point, it seems to me vital that the contested limits of these boundaries be acknowledged from an array of different vantage points from across the science–media nexus. Precisely what science *is*, and what it is *not*, is anything but straightforward, as even a cursory glance at, say, a daily newspaper item about '**mad cow disease**' or a television documentary about **global warming** will immediately make apparent. Some might argue that science, like beauty, is very much in the eye of the beholder. In any case, to think of science and the media as separate worlds in constant collision with one another may be advantageous in some ways, not least with regard to identifying key sources of tension in public debates, but should not be understood too literally. Much more appropriate, in my view, is a critical engagement with scientific and media discourses that accounts for the complex ways in which they each strive to engender certain preferred ways of talking about the nature of reality. Such an approach recognizes that the extent to which their respective truth-claims converge or, just as importantly, are made to diverge from one another, will have a profound impact on our sense of the world around us.

The line of inquiry I want to pursue in this book takes as its point of departure the thesis that efforts to discern what constitutes science must necessarily address the salience of media representations. As will be shown, ongoing debates about the ways in which the news and entertainment media represent scientific issues, and how their audiences are influenced as a result, are being constantly extended and enriched by new research from a wide array of disciplinary positions. It needs to be acknowledged here at the outset, however, that taken as a whole this research is still in a fairly rudimentary stage of conceptual and methodological development. Indeed, as Cooter and Pumfrey (1994) point out, surprisingly little has been written on science in popular culture over the years:

> Still shrouded in obscurity are the effects of even the most obvious mechanisms for the transmission of scientific knowledge and culture: the popular press, radio and television, to say nothing of science texts, museums, school curricula, and the overtly propagandist productions of the science lobby itself. From coffee houses to comic books and chemistry sets, from pulpits to pubs and picture palaces, from amateur clubs to advertising companies, from Science Parks to Jurassic Park, our ignorance both of the low drama and the high art of science's diffusion and modes of popular production and reproduction is staggering.
> (Cooter and Pumfrey 1994: 237)

Despite the relatively inchoate nature of research in this area, though, Cooter and Pumfrey are able to point to a number of significant recent developments, not least what they regard as a promising trend towards examining science as culturally situated. Both weak formulations ('science in culture') and stronger ones ('science as culture') share a recognition of 'how the shape and success of the sciences' are embedded in a complex array of social relations. These social relations link different scientific communities, in turn, with 'various allies, audiences, publics, consumers and reproducers; with powerful élites who bestow legitimacy and material support; and also with "lower" social groups whose willingness (or resistance) to engage with science is an equally important determinant of scientific culture' (Cooter and Pumfrey 1994: 240). It is the latter type of social relations, as Cooter and Pumfrey observe, that is frequently addressed by researchers as a matter of 'popularizing' science with the public.

Science and the public

Impassioned debates over how best to 'popularize' science are hardly new, and yet there appears to be a growing sense of urgency on the part of those who seek to speak on behalf of scientific inquiry today. Scientists themselves, as Dunbar (1995) points out, have seldom stopped for very long to query just what it is that characterizes their endeavour, and why it is important for society. 'Being pragmatic people,' he observes, 'they have simply got on and done it' (Dunbar 1995: 12).

Such a devotion to resolving the most pressing problems at hand clearly has its rewards, but it is also not without its limitations. Arguably one of the most significant of the latter is the extent to which this kind of pragmatism has hindered the development of a greater self-reflexivity among scientists about their collective responsibility to explain the nature of science to 'ordinary' or 'lay' members of the public. While scientists are usually quick off the mark to identify what they regard as unfortunate misconceptions in public debates about the aims and objectives of scientific inquiry, they are typically reluctant to engage in the task of clearing up these same misconceptions. Increasingly in the eyes of some in the scientific community, however, this reticence is becoming more and more difficult to justify. At issue, in their view, is the fear that support for science among members of the public is in a state of decline, especially where young people are concerned. Efforts to support this claim regularly rely on evidence gleaned from opinion polls and surveys attempting to measure the public's understanding of science, the results of which – in Western countries – often appear to show a steady increase in anti-science attitudes. Others dispute this interpretation, but in either case much is made of the corresponding decrease in the numbers of students enrolled on science courses at both school and university levels. This apparent trend is regarded by some commentators as having alarming implications for the next generation of science teachers and researchers, as well as for the future quality of modern society's industrial and technological development more generally.

Attributions of blame for this apparent disillusionment – if not outright antipathy – with science are being laid at the doorstep of a number of different possible sources by pro-science advocates. Typically singled out for criticism, for example, are voices from the arts, humanities and social sciences, many of which are regarded as exemplifying the worst sorts of tensions illuminated so sharply in C.P. Snow's (1965) *The Two Cultures* of several decades ago. Particularly worthy of censure today, in the opinion of these critics, are those social and cultural theorists who subscribe to the philosophical tenets of 'postmodernism' as a distinct intellectual position. Postmodernists, they argue, are deeply misguided in their questioning of the universality of reason, especially when they challenge the precept that there exists an external reality to be rationally detected through scientific modes of investigation. For these critics, the postmodernist dismissal of science as merely one way to understand the world, a 'language game' intrinsically no better or worse than others, is nothing short of contemptuous. Hence their outrage at what they perceive to be the growing influence of postmodernist attacks on science, not only in the educational system but increasingly in the media (see also McGuigan 1999). The failure of postmodernists to engage with science in an adequate manner, these critics insist, threatens to turn public scepticism about traditional standards of truth into bleak, nihilistic forms of alienation.

A further cause for concern among pro-science advocates is the challenge to science being mounted by individuals and groups anxious to resist its ascendancy in modern society. These voices reject the moral relativism of some postmodern definitions of scientific inquiry while, at the same time, refusing to countenance the claim that science is morally neutral. Such is the line of argument Appleyard (1992) outlines in his book *Understanding the Present*. 'Scientists need to be observed and criticized more than any other members of society,' he writes, 'not just because of the horrors that might emerge from their laboratories, but also because of the necessity for making them as morally and philosophically answerable as the rest of us' (1992: xiii). In taking exception with how science, in his view, supplants religion and culture, Appleyard contends that it is 'spiritually corrosive, burning away ancient authorities and traditions' with devastating effectiveness. It is inevitable, he argues, that scientists 'take on the mantle of wizards, sorcerers and witchdoctors', thereby ensuring that their 'miracle cures' become 'our spells, their experiments our rituals' (1992: 9). It is thus the very 'terrifying success' of science which needs to be resisted for fear that its cold logics will culminate in the destruction of fundamental human values, and with them the very sense of self that endows life with meaning, purpose and emotional fulfilment. In calling for the end of the 'rule of science', then, Appleyard is issuing a bold challenge to those pro-science advocates claiming to be able to find in science the basis for a spiritually enriched appreciation of the world around them.

One such advocate willing to respond to this challenge is Dawkins (1998), an evolutionary scientist. Interestingly, the title of his book, *Unweaving the Rainbow*, refers to a line in a poem by Keats, 'who believed that Newton had destroyed all the poetry of the rainbow by reducing it to the prismatic colours' (Dawkins 1998: xii). Keats could not have been more wrong, Dawkins argues, not only because the poetry of the rainbow survives Newton's explanation of its properties, but also because science is – or ought to be – the stimulus for great poetry. Dawkins (1998: 26) is determined to counter the perception that 'science is poetry's killjoy, dry and cold, cheerless, overbearing and lacking in everything that a young Romantic might desire'. Indeed, he goes so far as to offer the 'untestable speculation' that Keats 'might have been an even better poet if he had gone to science for some of his inspiration' (1998: 27). More to the point, though, Dawkins considers it vital to recognize that the 'spirit of wonder' which motivates great poets similarly moves great scientists. And besides, as he points out, 'Newton's dissection of the rainbow into light of different wavelengths led on to Maxwell's theory of electromagnetism and thence to Einstein's theory of special relativity' (1998: 42).

It is precisely this 'spirit of wonder' as it informs a scientific sensibility which is under threat, in Dawkins's opinion, by the various forms of **pseudoscience** regularly being presented in the media. Particularly disturbing are those forms of popular culture which proclaim for themselves – either explicitly or, more typically, implicitly – a scientific status that is completely unwarranted. For example, the 'meaningless pap' of astrology is offensive, in his view, 'especially in the face of the real universe as revealed by astronomy', but also because of the 'facile and potentially damaging way in which astrologers divide humans into 12 categories' or 'signs' under the horoscope's star system (Dawkins 1998: 115, 118). Considering one's horoscope in a newspaper, for some readers a lighthearted form of amusement, can have unfortunate consequences for the vulnerable anxious to follow the advice on offer. Similarly, he cites the media's 'obsession' with the paranormal, taking strong exception with 'second-rate conjurors masquerading as psychics and clairvoyants' (1998: 115). From there it is arguably a short step to believing in telepathy and magic, or even entertaining the possibility of spirits and hobgoblins existing. 'Disturbed people recount their fantasies of ghosts and poltergeists,' he writes,

> But instead of sending them off to a good psychiatrist, television producers eagerly sign them up and then hire actors to perform dramatic reconstructions of their delusions – with predictable effects on the credulity of large audiences.
>
> (Dawkins 1998: 129)

And, it may be said, with a corresponding impact on what comes to be understood as 'science' as a result.

Turning to television's representation of scientific inquiry itself, Dawkins singles out for criticism the popular fiction series *The X-Files*. Each episode, he points out, revolves around a mystery confronted by two Federal Bureau of Investigation (FBI) agents: 'One of the two, Scully, favours a rational, scientific explanation; the other agent, Mulder, goes for an explanation which either is supernatural or, at very least, glorifies the inexplicable' (Dawkins 1998: 28). Both agents struggle to resolve the mystery by bringing to bear insights garnered from their respective positions, a narrative device which for some can be justified as harmless fiction. Dawkins disagrees, insisting that the 'problem with *The X-Files* is that routinely, relentlessly, the supernatural explanation, or at least the Mulder end of the spectrum, usually turns out to be the answer' (1998: 28). The defence of the programme as harmless fiction is thus a hollow one, in his view, because it 'systematically purveys an anti-rational view of the world which, by virtue of its recurrent persistence, is insidious' (1998: 28). To the extent that *The X-Files* refuses to portray the world as an orderly place, where certain laws of nature apply, it fails Dawkins's test of good science fiction. 'Science', he contends, 'allows mystery but not magic, strangeness beyond wild imagining but no spells or witchery, no cheap and easy miracles' (1998: 29).

We shall return to *The X-Files* at a later point in this book's discussion, but here it is important to observe how Dawkins's intervention similarly encompasses efforts to 'dumb down' science in the name of making it more publicly accessible. Dawkins sees in the 'public understanding of science' movement in countries like Britain and the US, for example, an anxiety among scientists to be loved by members of the public (see also Chapter 3). As he writes:

> Funny hats and larky voices proclaim that science is fun, fun, fun. Whacky 'personalities' perform explosions and funky tricks. I recently attended a briefing session where scientists were urged to put on events in shopping malls designed to lure people into the joys of science. The speaker advised us to do nothing that might conceivably be seen as a turn-off. Always make your science 'relevant' to ordinary people's lives, to what goes on in their own kitchen or bathroom. Where possible, choose experimental materials that your audience can eat at the end [. . .] The very word science is best avoided, we were told, because 'ordinary people' see it as threatening.
>
> (Dawkins 1998: 22)

Calculated efforts to achieve this 'dumbing down' such as these ones may indeed attract the interest of the public, he argues, but they can all too easily become condescending and patronizing or, even worse, turn into a kind of 'populist whoring that defiles the wonder of science'. His choice of the latter phrase, in my view an inappropriate one, is in any case indicative of the passion with which he declares that 'we shouldn't need whacky personalities and fun explosions to persuade us of the value of a life spent finding out why we have life in the first place' (Dawkins 1998: 23). Thus while disputes about the 'dumbing down' of science have been around for a long time, Dawkins's voice is part of a growing chorus from within the scientific community demanding a counter-attack of sorts against what they perceive to be an increasingly dangerous misrepresentation of science in the media.

Risky science, uncertain media

Still, is it not the case, one might be tempted to ask in light of these kinds of criticisms, that most images of science in the media are positive ones? Numerous examples promptly come to mind, whether from the daily news, advertisements, television documentaries or the like, where science is seen to be providing answers, solving problems, and making the world a better place. A science programme devoted to helping us to understand the origins of the universe may very well be interrupted, in turn, by a commercial telling us that thanks to the use of science, a particular brand of shampoo is much more effective at cleaning hair than another one.

Anecdotal evidence aside, one need not proceed very far down this line of argument to reach the conclusion that broad, sweeping generalizations about media images of science are of little heuristic value. Sufficient care must be taken to attend to the complex, frequently contradictory ways in which media representations are produced, distributed and negotiated across the public and private realms of society. Important here is a context-sensitive approach to any bold claim – whether positive or negative – about how people are being influenced (or not) by these representations in the situational circumstances of their everyday lives. Indeed, as will be argued in later chapters of this book, this imperative is particularly significant where issues of '**risk**' are concerned. Just as one cannot safely extrapolate from opinion polls to characterize the 'public understanding of science' in a narrow sense, in my view, one should similarly avoid the assumption that public perceptions of risks, threats and hazards can be ascertained by examining media content alone. At the same time, however, one need not invoke a language of causative media 'effects' or 'impacts' to acknowledge the absolutely crucial ways in which people draw upon media representations to help them make sense of different scientific controversies. Is nuclear power safe? Am I in danger of contracting **HIV/AIDS**? Will I get the human variant of 'mad cow disease' if I eat a beefburger? What about **genetically modified (GM) crops** or **foods**, are they too risky? Should I be worried about human **cloning**? These everyday questions, posed among a myriad of related ones, go to the heart of this book's elucidation of the changing relationships between media, risk and science.

It is my aim in this book to contribute to the development of a conceptual framework for the purpose of engaging with several pressing debates in this exciting area of inquiry. Not surprisingly, the task of weaving together insights from multiple strands of research emergent across diverse disciplines has not been easy or straightforward, and the end result is not comprehensive in its purview. It could not be otherwise, of course, and therein lay the challenge. My approach has been necessarily selective given the limitations of space, leading me to dwell on aspects of these debates which I hope will be of particular interest to the reader. Far from being yet another salvo in the so-called **'science wars'** (or a response to the **Sokal affair**, which surely has lost whatever ardour it possessed for some by now), this book seeks to facilitate new connections being made across otherwise disparate research literatures. In elaborating a conceptual framework, it strives to help improve the conditions for an enriched dialogue, one that may inform future explorations of these issues well beyond traditional disciplinary boundaries. Accordingly, a brief overview of the different chapters is as follows:

- 'Today's science fiction', as Stephen Hawking (1995: xiii) has observed, 'is often tomorrow's science fact.' Bearing this point in mind, Chapter 2 provides an evaluative assessment of different representations of science and scientists in science fiction. The emergence of science fiction as a distinct genre is examined, with special attention given to the literary works of Jules Verne and H.G. Wells, respectively, as well as to the formative role played by 'pulp' magazines. The prospect of interplanetary travel, arguably the most celebrated of science fiction themes, is discussed with special reference to the different *Star Trek* series. Overall, this chapter's focus on the science of science fiction shows how it speaks directly to people's concerns, fears, anxieties and desires, encouraging them to work through the possible implications of different scenarios while, at the same time, promising to keep them entertained in the process.
- Informing Chapter 3's discussion is a key question, namely what counts as 'science' in popular culture today? The question is shown to be a pressing one, namely because of the growing concern among some scientists that the boundaries protecting scientific facts from pseudo-science speculation are crumbling – not least, they contend, because of the irresponsible ways in which distinctions between the two are represented by the media. This chapter thus takes as its point of departure the thesis that popular cultural images of science and scientists – and here *The X-Files* is shown to be an important case in point – are worthy of sustained investigation. Indeed, it will be argued that these images are crucial factors which need to be carefully accounted for in any critical effort to rethink the central tenets of the 'public understanding of science' debate, especially with regard to **scientific literacy**. In exploring some of the implications of this thesis for this debate, science museums and centres receive particular attention.
- In Chapter 4, the focus shifts to science journalism. In examining the factors shaping how news organizations report on the world of science, a

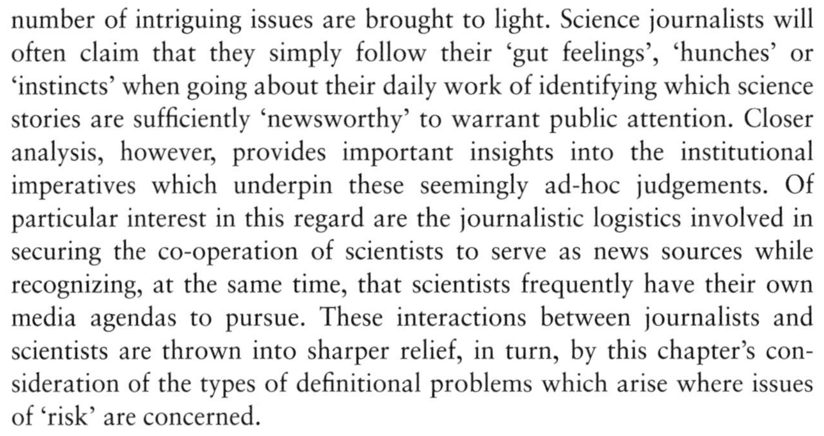

number of intriguing issues are brought to light. Science journalists will often claim that they simply follow their 'gut feelings', 'hunches' or 'instincts' when going about their daily work of identifying which science stories are sufficiently 'newsworthy' to warrant public attention. Closer analysis, however, provides important insights into the institutional imperatives which underpin these seemingly ad-hoc judgements. Of particular interest in this regard are the journalistic logistics involved in securing the co-operation of scientists to serve as news sources while recognizing, at the same time, that scientists frequently have their own media agendas to pursue. These interactions between journalists and scientists are thrown into sharper relief, in turn, by this chapter's consideration of the types of definitional problems which arise where issues of 'risk' are concerned.

- The pressing need to better understand how the media represent risk forms the basis of Chapter 5's discussion, in this case with respect to environmental risks. In the first instance, various 'myths of nature' are examined, in part to render problematic what counts as 'the environment' as a social problem. Attention then turns to Ulrich Beck's (1992a) conception of the **risk society** so as to help to identify, in turn, how the news media shape the 'relations of definition' (Beck 2000) framing public discourses about environmental risks, threats and hazards. In order to extend further this line of argument, an evaluative assessment is made of a variety of different studies concerned with news coverage of environmental issues. Particular attention is devoted to nuclear technologies in this context, before the chapter moves to highlight a range of pertinent insights provided by environmental journalists themselves. It is argued that while the news media are playing a crucial role in sustaining the imperatives of 'expert' risk assessment, they are also creating spaces, albeit under severe constraints, for counter-definitions to emerge.
- Chapter 6 examines how the media have reported on one of the worst epidemics in the history of humankind, namely HIV/AIDS. Currently about 15,000 new cases are being identified around the world each day, and these numbers are rapidly escalating – the crisis is threatening to quickly overwhelm efforts to control it. Critical research on media portrayals of HIV/AIDS suggests that much of the pertinent news coverage has contributed to the creation of the hostile climate – typically characterized as a 'moral panic' – around the disease. While some journalists have refused to succumb to sensationalist reporting, far too many others have produced news accounts which effectively dehumanize people living with the condition. This chapter focuses on the early years of the news coverage in Britain and the US, showing how the primary definitions of

the crisis were set in motion, and the difficulties groups have had since in seeking to disrupt, challenge and reverse their ideological purchase.

- In order to elucidate several key issues surrounding food risks, Chapter 7 takes as its focus news media coverage of food scares. 'Eating', as Caplan (2000: 187) observes, 'has become a risky business.' The discussion begins with an examination of the anatomy of a food scare, in general, before proceeding to examine two case studies for illustrative purposes. Specifically, the news reporting of two ongoing crises is subjected to close scrutiny: first, the identification of **BSE** or 'mad cow disease' in Britain's cattle, and the ensuing 'beef crisis' which now threatens to spread around the world. The second crisis revolves around public fears about genetically modified (GM) crops and foods, where a language of 'Frankenfoods' and 'killer tomatoes' has permeated public culture. In a concluding section which considers what lessons are being learned from these crises, questions are raised about how best to improve the ways in which the media represent images of science so as to further enhance democratic accountability over public, corporate and scientific decision-makers.
- What does it mean to be human? This is the question which launches Chapter 8's inquiry. Attention first turns to the tale of *Frankenstein* due to its status – as Turney (1998: 3) argues – as the 'governing myth of modern biology'. The salience of this myth, and the extent to which it is drawn upon as a rhetorical resource to both advance and challenge certain formulations of science, receives particular attention. The respective figures of the **robot**, **android** and **cyborg** are then examined so as to discern how familiar boundaries between 'human self' and 'mechanical others' have been recast from alternative perspectives within science fiction. Next, the brave new world of **IVF** and 'test-tube babies', along with genetically engineered 'designer children', is centred for discussion. Anxieties about humans being taken over by robots are being replaced, it seems, with fears that humans will be genetically engineered into becoming robot-like themselves. The final section considers the prospect of human cloning. Whether science is out of control (*Frankenstein*) or in control (*Brave New World*), the possible implications of cloning people – that is, creating babies that are genetic replicas of adults – could hardly be more profound for humankind.

2 SCIENCE FICTIONS

> Fiction can be defined as a coherent tissue of lies. When science provides the coherence, we call it science fiction. What makes science fiction so exciting is that contemporary science, while full of useful rules and formulas masquerading as laws, is a work in progress. Instead of constraining creativity with full-blown dogmas, science erects signposts saying, Look here! Go there! Danger: sharp curves ahead!
>
> (Gerald Jonas, science fiction reviewer, *The New York Times*)

'Those who have never seen a living Martian', we are warned, 'can scarcely imagine the strange horror of its appearance.' As the Martian invader slowly emerges from its space capsule, an all too vivid description proceeds to unfold:

> The peculiar V-shaped mouth with its pointed upper lip, the absence of brow ridges, the absence of a chin beneath the wedge-like lower lip, the incessant quivering of this mouth, the Gorgon groups of tentacles, the tumultuous breathing of the lungs in a strange atmosphere, the evident heaviness and painfulness of movement due to the greater gravitational energy of the earth – above all, the extraordinary intensity of the immense eyes – were at once vital, intense, inhuman, crippled and monstrous. There was something fungoid in the oily brown skin, something in the clumsy deliberation of the tedious movements unspeakably nasty. Even at this first encounter, this first glimpse, I was overcome with disgust and dread.
>
> (Wells [1898] 1993: 19–20)

This account belongs, of course, to the narrator of H.G. Wells's novel *The War of the Worlds*, published in 1898. This remarkable novel, subsequently adapted into a radio programme in 1938 (all too memorably so, as its broadcast sparked widespread panic among many listeners) and later several films and a television series in the US, appeared decades before the term 'science fiction' gained popular currency. And yet, for many people, *The War of the Worlds* neatly encapsulates a number of the features that make science fiction such an engaging genre. Among them is the way the novel transports us into a different world, one where humans discover not only that they are not alone in the universe, but also that they are scientifically and technologically inferior to their 'alien' rivals. Even today Wells's depiction of genocidal invaders from Mars continues to resonate in popular imagery of the planet, not least in news reports about scientific debates concerning whether some form of organic life has ever existed on its surface. The prospect of life on Mars, thanks in part to this novel, cannot be thought of as purely a matter of scientific interest. 'Though one cannot blame Wells for all the later excesses of interplanetary warfare,' science fiction writer Arthur C. Clarke (1993: xxxii) points out, 'perhaps he merits some criticism for propagating the creed that anything alien is likely to be horrible.'

In tracing the evolution of science fiction as a distinct literary genre, historians tend to have their own preferred contender for the key figure or significant event which most clearly signals what to their mind is the essential starting point. Opinions vary quite considerably with regard to which specific criteria should be invoked when making such a judgement, with some making the case for writers as distant as Lucian of Samosata, who wrote tales of heroes sprouting wings and flying to the Moon in the second century AD. In contrast, I tend to side with those who deem Mary Shelley's *Frankenstein*, first published in London in 1818, to be the most notable precursor of what would eventually become a full-fledged genre only in the twentieth century. Crucially, the plausibility of this novel's account of how life may be created in the laboratory rests on Dr Victor Frankenstein's application of certain (imagined) scientific facts and principles with his 'instruments of life', as opposed to issuing an appeal to a divine or transcendental power. It is this distinction between the natural and the supernatural, as several commentators have pointed out, that marks the novel's decisive break from earlier types of literary fantasy which revolved around the latter. If Shelley, as Turney (1998: 2–3) argues, 'did not, in any plausible usage, offer predictions about the future of science', she nevertheless did 'identify concerns which go to the heart of our response to science.' Indeed, Turney suggests that this 'story about finding the secret of life became one of the most important myths of modernity.'

This chapter is primarily concerned with exploring images of science in science fiction more broadly, so we shall return to the importance of the Frankenstein myth for modern biology in Chapter 8. Science fiction is, of course, a widely enjoyed form of popular entertainment with highly dedicated fans, and as such has been the subject of intense scrutiny by academic commentators from a wide variety of perspectives. Our focus here will be restricted to exploring some of the ways in which science fiction gives public expression to scientific issues. This line of inquiry necessarily takes issue with the argument that science fiction may be safely dismissed as being irrelevant to 'real' science, that its imagery merely reflects – either accurately or in a distorted manner – the objective truth of scientific laws. In contrast, this chapter begins the work, to be further elaborated upon in subsequent chapters, of assessing the implications these representations may have for how people – including scientists – make sense of scientific controversies in the context of their everyday lives. Accordingly, our attention turns in the first instance to tracing the literary ancestry of the term 'science fiction' so as to gain a sense of how its emergence and development as a recognizable genre of fiction has been shaped by discourses of science.

Science in science fiction

Efforts undertaken to formally define 'science fiction' are confronted from the outset with the obvious problem that the term itself is something of an oxymoron. To hold 'science' in conjunction with 'fiction' is certain to call forth highly contradictory associations, to say the least. The struggle to keep these two elements in some sort of balance typifies much of what makes the whole genre relatively distinct from neighbouring ones, such as 'horror' or 'fantasy'. Still, it quickly becomes apparent upon closer inspection that the boundaries demarcating the outside limits of this genre are constantly being drawn and redrawn across disparate media contexts. As the criteria informing judgements about what should count as science fiction are always changing, some members of the science fiction community favour using the label 'sf' instead. Not only can sf be made to stand in for science fiction, but also it can be retranslated in line with more specific personal tastes and preferences, such as 'science fable', 'speculative fiction', 'scientific fantasy' or 'science-in-fiction', among other alternatives. Certainly the one term likely to attract derision from across the breadth of this community is 'sci-fi', due to its perceived correlation with the 'low grade pulp sf' enjoyed by those judged not to be true fans of the genre.

Given these kinds of semantic tensions, it is hardly surprising that the task of identifying a satisfactory definition of science fiction is a challenging one. Indeed, some would go so far as to suggest that any such attempt is bound to be futile, and possibly counterproductive if it proves too narrow or restrictive. Even a broadly sketched definition would need to encompass remarkably diverse communities of writers and their readers, listeners or viewers, as well as the critics, reviewers and those who have some investment (literally for those with a financial interest) in determining its 'origins' or status as a genre. In any case, and regardless of the chosen definition, for a specific novel, film or television programme to be accepted as belonging to the realm of science fiction it will need to incorporate (to some extent) a range of interrelated conventions, narrative codes and stylistic devices. These characteristic features, so familiar that they are usually difficult to deconstruct or render problematic, help to orient the intended audiences to make sense of the science fiction text in particular ways – just as a different array of expectations are associated with, say, westerns, romances or detective stories. Science fiction is recognizable as such, it follows, because it respects certain kinds of rules.

To begin the work of pinpointing the principal features of science fiction for our purposes here, it is advantageous to briefly trace the emergence of the genre in historical terms. Several historians of science fiction maintain that the actual phrase 'science fiction' first appeared in 1851. The occasion was a treatise on the poetry of science, titled *A Little Earnest Book upon a Great Old Subject*, prepared by the English writer William Wilson. A key passage from Chapter 10 reads:

> All known Sciences contain within themselves Worlds of exquisite Poetry, and the more the general mind becomes familiarised with the ever-varying interest and fascinations connected with their Study, the more rapid will become the diffusion and the rise of Science. Science is a holy devotion, and the pursuits and the results attained are alike glorius [. . .]
>
> Campbell [the Scottish poet Thomas Campbell] says that 'Fiction in Poetry is not the reverse of truth, but her soft and enchanting resemblance.' Now this applies especially to Science-Fiction, in which the revealed truths of Science may be given, interwoven with a pleasing story which may itself be poetical and *true* – thus circulating a knowledge of the Poetry of Science, clothed in a garb of the Poetry of Life.
>
> (cited in Evans 1999: 180)

Wilson's treatise is interesting at a number of different levels, but particularly so because of its emphasis on the potential of science fiction for helping to put into public circulation ostensibly rational explanations for the 'revealed truths of science'. Such an emphasis in science fiction is in marked contrast with other types of fiction being published around this time, especially works of fantasy – imaginings of the impossible – intended strictly for purposes of entertaining the reader. It is in these related categories, including those of 'speculative' fiction but especially 'scientific romances', that narratives by authors such as Alexandre Dumas, Nathaniel Hawthorne, Victor Hugo, Fitz-James O'Brien, Edgar Allan Poe and Mark Twain are regularly placed.

In this context it is worth observing that French writer Jules Verne's short story 'A Balloon Journey', arguably the first of his '*voyages extraordinaires*', also appeared in 1851 (his first scientific adventure novel, *Five Weeks in a Balloon*, was published in 1863). Considered by many commentators today to be the 'father of science fiction', Verne's stories consistently demonstrated his personal commitment to extending a pedagogical agenda. 'Verne was essentially an autodidact in scientific matters,' observes Lambourne (1999: 147), 'showing the common enthusiasm of the self-taught for passing on newly acquired knowledge, especially in the many books he wrote for the publisher Hetzel, with a juvenile audience in mind'. Some biographical accounts suggest that it was Verne's introduction the year before to explorer and adventurer Jacques Arago, brother of the scientist François Arago, that encouraged him to engage in a sustained private study of geography, physics and mathematics. In literary terms, Verne took inspiration from sources such as the writings of Poe, a point he readily acknowledged. Over the years he would interweave scientific observations into fictional narratives rich with prophetic visions about technical marvels to wide public acclaim. His major books included such titles as: *A Journey to the Centre of the Earth* (1864; where travellers enter the earth's crust through a dormant volcano); *From the Earth to the Moon* (1865; where space travellers are launched into orbit courtesy of a giant cannon); *20,000 Leagues Under the Sea* (1870; where Captain Nemo's submarine 'Nautilus' appears); and *The Clipper of the Clouds* (1886; where a heavier-than-air flying machine is held aloft by 74 vertical propellers).

Nevertheless, it was not Verne's explicit aim to produce science fiction, a term that would not gain popular currency for years to come in any case. The vast majority of the stories he wrote (some penned with his son Michel) were tales of adventure, many revolving around imaginary journeys to exotic places, where science is drawn upon primarily to enhance narrative plausibility. Few would object with the assertion, however, that Verne's writings made a vital contribution to laying the foundations of the science fiction genre. His determination to avoid the excesses of what he termed 'irresponsible' scientific speculation meant that his stories characteristically stayed largely within the boundaries of scientific knowledge and certain prototypes of new technologies emerging at the time. In the words of Angenot (1979: 31), a literary critic, 'Verne is paradoxically a utopianist without an alternative society; he is the last SF writer who believes in industrialist euphoria, even if some pessimism overshadows his last books.' Regarding the latter novels, Angenot would presumably have in mind the dystopian work *Paris in the Twenty-First Century*, among others, but his point that Verne is the 'last happy utopianist' is a fair one. If at times Verne's writing has been found wanting in terms of its literary virtues (and here the translation from the French is often to blame), it was extraordinarily successful in capturing his reading public's imagination concerning the wondrous promise of scientific progress.

Easily the most noteworthy of the other figures at the time whose writing can be hailed as a precursor of science fiction is the British writer H.G. (Herbert George) Wells. Typically singled out for discussion by historians of science fiction are books such as *The Time Machine* (1895), *The Island of Doctor Moreau* (1896), *The Invisible Man* (1897), *The War of the Worlds* (1898), *The First Men in the Moon* (1901) and *The Shape of Things to Come* (1933). Despite being widely credited with revolutionizing the development of this still inchoate genre, Wells himself claimed to regard his own writing technique as little more than 'an ingenious use of scientific patter' (cited in Alkon 1994: 43). In contrast with Verne's didactic approach to science, he sought to ensure that his narratives would 'reflect upon contemporary political and social discussions' so as to extend his personal commitment to advancing pointed criticism of social trends. Such an objective meant that precise details about the scientific technologies being introduced were frequently ignored in favour of a commentary on their broader political and philosophical implications for society.

Interviewed in *TP's Weekly* in 1903, Verne dismissed Wells's (1901) *The First Men in the Moon* by speaking directly to this kind of tension:

> It is very curious, and, I will add, very English. But I do not see the possibility of comparison between his work and mine . . . it occurs to me that his stories do not repose on very scientific bases. No, there is no rapport between his work and mine. I make use of physics. He invents. I go to the moon in a cannonball, discharged from a cannon. Here there is no invention. He goes to Mars in an airship, which he constructs of a material which does away with the law of gravitation. That's all very well but show me this metal. Let him produce it.
>
> (cited in afterword to Wells [1901] 2001)

It is obvious to note, of course, that Verne's vision of people being launched into space via a cannon was hardly any more practical than Wells's preferred scenario, even if he did include detailed calculations of escape velocity and transit times. Still, it is the case that Wells, while boldly moving beyond the limitations of the scientific romance in narrative terms, did not share Verne's deeply felt reverence for scientific facts and figures. Indeed, the actual scientific premises underlying Wells's stories were frequently less than credible, some even being absurd. That said, the opposite was sometimes true too, as certain insights appeared to be indicative of the 'raw material' he had acquired through his formal training in biology at the Normal School of Science (later the Royal College) London. There he studied under such scholars as T.H. Huxley, a passionate advocate of Darwin's evolutionary theory.

It was while working as a teacher of science, living in near poverty, that Wells began to write for publication. His first science fiction story (if not recognized as such at the time) was *The Time Machine*, written when he was almost 30, and it garnered immediate success with the public and critics alike upon its appearance in 1895. The short novel depicts the experiences of a time traveller who is propelled forward to the year 802701, where he encounters a society divided between two classes: the Morlocks, who live and work beneath the earth's surface, and the Eloi, who enjoy lives of decadence above ground in idyllic comfort. Serving as a social allegory, the novel gives expression to Wells's views on Western society in a manner at once critical and engaging. In each of the novels which followed over the next few years, he sought to bring to life abstract scientific postulates in excitingly unexpected ways, and always in conjunction with an underlying critique of society (and, at times, humankind itself). None of the early novels is more successful in this regard, at least in my view, than *The War of the Worlds*, which appeared in 1898, having been published in serial form the previous year.

Wells's fifth novel, *The War of the Worlds* describes in vivid terms the arrival of Martian invaders on earth. Despite the concerted effort of the military, virtually all human opposition is vaporized in flashes of 'luminous green smoke' directed from the heat-rays of killing machines (those few humans spared from the onslaught having been collected in metal traps). The Martians promptly proceed to destroy a rapidly evacuated London before succumbing, in the end, to common bacterial microbes that prove lethal to their fragile bodies. Still, the novel is much more than a harrowing tale of disaster. The opening paragraph sets an introspective tone, carefully sustained throughout:

> No one would have believed in the last years of the nineteenth century that this world was being watched keenly and closely by intelligences greater than man's and yet as mortal as his own; that as men busied themselves about their various concerns they were scrutinised and studied, perhaps almost as narrowly as a man with a microscope might

scrutinise the transient creatures that swarm and multiply in a drop of water. With infinite complacency men went to and fro over this globe about their little affairs, serene in their assurance of their empire over matter. It is possible that the infusoria under the microscope do the same. No one gave a thought to the older worlds of space as sources of human danger, or thought of them only to dismiss the idea of life upon them as impossible or improbable. It is curious to recall some of the mental habits of those departed days. At most terrestrial men fancied there might be other men upon Mars, perhaps inferior to themselves and ready to welcome a missionary enterprise. Yet across the gulf of space, minds that are to our minds as ours are to those of the beasts that perish, intellects vast and cool and unsympathetic, regarded this earth with envious eyes, and slowly and surely drew their plans against us. And early in the twentieth century came the great disillusionment.

(Wells [1898] 1993: 5)

This description of the invading Martians, together with the kinds of details provided in passages like the one appearing at the start of this chapter above, strikes a chord that is simultaneously strange and familiar. That is to say, Wells's Martians, as Aldiss (1973: 119) argues, 'are creatures not of horror but terror; they spring from a sophisticated acknowledgement that they are all part of us, of our flesh.' Evil is not simply externalized or projected upon creatures alien to us, rather the story's narrator implores the reader to recognize in humanity itself the propensity for malfeasance. The imperialist drive of the Martians is consistent with how humans have treated one another throughout history and, moreover, may be foreshadowing the eventuality of human colonization of distant planets. In any case, as O'Brien (2000: 50) maintains, this novel 'was to have a profound effect on the British science fiction tradition: the first of many dark, dystopian and frequently tortured visions of armageddon' (see also King and Krzywinska 2000; Roberts 2000).

Readers of Wells today are likely to be struck by the remarkable power of some of his novels' prophecies. He displayed a talent for being able to anticipate scientific developments, such as the splitting of the atom in *The World Set Free* (1914) decades before it took place. The Hungarian physicist Leo Szilard, who played an instrumental role in the development of the atomic weapons programme in the US, would later recall:

In 1932, while I was still in Berlin, I read a book by H.G. Wells. It was called *The World Set Free*. This book was written in 1913, one year before the World War, and H.G. Wells describes the discovery of artificial radioactivity and puts it in the year 1933, the year in which it

> actually occurred. He then proceeds to describe the liberation of atomic energy on a large scale for industrial purposes, the development of atomic bombs, and a world war which was apparently fought by an alliance of England, France, and perhaps including America, against Germany and Austria, the powers located in the central part of Europe.
> (cited in Canaday 2000: 3)

Having taken his inspiration from Wells's novel, Szilard turned his attention to considering the problem of how a nuclear chain reaction might be sustained, eventually patenting an approach in the name of the British Admiralty. He had become convinced that 'the forecast of the writers may prove to be more accurate than the forecast of the scientists', a possibility with profound implications: 'Knowing what [the possibility of a chain reaction] would mean – and I knew it because I had read H.G. Wells – I did not want this patent to become public' (cited in Canaday 2000: 4). In due course Szilard would draft a letter for Albert Einstein to send to US President Franklin Roosevelt, warning him of the possible construction of 'extremely powerful bombs of a new type'. Roosevelt's response set in motion a committee that would establish, in turn, the Manhattan Project, the top secret wartime project to build an atomic bomb. The familiar distinction between scientific 'fiction' and 'fact' was thus decisively recast, as in Szilard's view Wells's portrayal of atomic warfare had become 'suddenly real'.

Amazing tales of super-science

The actual term 'science fiction' did not achieve widespread use until 1929, long after Wilson's description of it in 1851 had been forgotten. Credit for introducing it to the public lexicon is customarily given to Hugo Gernsback, owner and editor of such pulp magazines as *Amazing Stories: The Magazine of Scientifiction* in the United States. The word 'pulp', incidentally, referred to the relatively cheap quality of the paper being used at the time, although others have regarded it (fairly or not) as a synonym for certain perceived shortcomings in the magazines' literary merits.

Many of the stories appearing in *Amazing Stories* took their inspiration from writers like Verne and Wells, together with that of Poe. Part of the reason for Gernsback's emphasis on their work was because of the financial savings to be gained for copyright clearance and related production costs due to the commercial reprint policy of the day. The first edition of *Amazing Stories* (April 1926), for example, had appeared on newsstands containing

Verne's 'Off on a Comet', Wells's 'The New Accelerator' and Poe's 'The Facts in the Case of M. Valdemar', among other stories. Reprints of the famous also helped to fill out pages in light of the relatively small number of newcomers to the field. At the same time, however, it is clear that Gernsback (1926: 3) was truly inspired by the ways in which these stories engendered, in his words, 'charming romance intermingled with scientific fact and prophetic vision'. It was this these three elements – romance to provide entertainment, science for purposes of education, and prophecy to offer inspiration – which underpinned the proclaimed genre. Gernsback's pedagogical agenda was stated in unambiguous terms in the first issue's editorial, an extract from which reads:

> It must be remembered that we live in an entirely new world. Two hundred years ago, stories of this kind were not possible. Science, through its various branches of mechanics, electricity, astronomy, etc., enters so intimately into all our lives today, and we are so much immersed in this science, that we have become rather prone to take new inventions and discoveries for granted. Our entire mode of living has changed with the present progress, and it is little wonder, therefore, that many fantastic situations – impossible 100 years ago – are brought about today. It is in these situations that the new romancers find their great inspiration.
>
> Not only do these amazing tales make tremendously interesting reading – they are also always instructive. They supply knowledge that we might not otherwise obtain – and they supply it in a very palatable form. For the best of these modern writers of scientifiction have the knack of imparting knowledge, and even inspiration, without once making us aware that we are being taught.
>
> (Gernsback 1926: 3)

Gernsback's use of 'we', unfortunately, was all too indicative of the times. The writers of the stories were almost always white men, and in their narratives women only rarely made an appearance (while 'alien species', as Ross (1991: 111) observes, 'were typically vilified in the most overtly racist ways'; see also Cartmell et al. 1999). The first story to appear by a female author was Clare Winger Harris's 'The Fate of the Poseidonia', which she submitted for a $500 prize best story contest featured in the December 1926 edition. Winning third place, she soon became a regular contributor (see also Hartwell 1996; Ackerman 1997; Ashley 2000).

In 1929, Gernsback launched a popular magazine called *Science Wonder Stories*. Having lost editorial control of *Amazing Stories* earlier that year, he had sought an alternative to the notion of 'scientifiction' (evidently another term of his coinage, but soon regarded as unwieldy) so as to better differentiate his new magazine in the marketplace. In using the term 'science fiction', Gernsback was of the view that he was simply attaching a new label to a genre already pretty much in existence. That said, however, *Science Wonder Stories* promptly declared its formal commitment to distinguishing the principal features of the genre. Gernsback's editorial policy statement for the first issue (June 1929) made clear the criteria for inclusion he intended to follow: 'It is the policy of *Science Wonder Stories* to publish only such stories that have their basis in scientific laws as we know them, or in the logical deduction of new laws from what we know' (cited in Parrinder 1980: 13). Moreover, his vision of the main rationale for science fiction is similarly worth quoting here:

> Not only is science fiction an idea of tremendous import, but it is to be an important factor in making the world a better place to live in, through educating the public to the possibilities of science and the influence of science on life . . . If every man, woman, boy and girl, could be induced to read science fiction right along, there would certainly be a great resulting benefit to the community . . . Science fiction would make people happier, give them a broader understanding of the world, make them more tolerant.
>
> (cited in James 1994: 8–9)

The title of the new magazine was shortened to *Wonder Stories* in June 1930, and continued to place great emphasis on depicting science as accurately as possible – a laudable objective, but one not always achieved. In addition to its stories, the magazine offered science quizzes, a 'Science Questions and Answers' section, and its list of 'Associate Editors' included the names of several luminaries from the world of science. To this day, some critics continue to lament Gernsback's formative role in the introduction of science fiction, particularly his determination to differentiate – some say isolate – the genre from other literary traditions. 'Nonetheless,' as Lambourne et al. (1990: 18) maintain, 'for good or for ill, Gernsback did help to establish the style of the American pulp science fiction market and thereby influenced enormously the subsequent development of the field.'

Throughout the 1930s, the pulp magazines were recognized as the genre's most successful exemplars. Despite the 'universal' language of science being popularized by the pulps, however, editors like Gernsback had long recognized that it was white male audiences (a sizeable share of which was made up of adolescents) who were most attracted to their stories. Not surprisingly, efforts were slowly made to extend the reach of the publications, as Aldiss (1973) observes in his historical account of the formation of the science fiction community:

> Gernsback soon discovered and made use of an active fandom, lads [sic] who read every word of every magazine with pious fervour and believed every word of editorial guff. These fans formed themselves into leagues and groups, issued their own amateur magazines or 'fanzines', and were generally a very vocal section of the readership. Many writers and editors later rose from their ranks. This particular factor of a devoted and enthusiastic readership is peculiar to science fiction, then and now.
>
> (Aldiss 1973: 216)

In time, these fans would take their desire to communicate with one another well beyond the letters columns of the magazines themselves. Gernsback, using his editorials in *Wonder Stories*, had proceeded to launch the first fan organization, the Science Fiction League (SFL), in 1934. While the League itself was short lived, several of its local chapters and their respective fanzines prospered independently, including several of those established in Britain (the first overseas chapter of the SFL having been formed in Leeds in 1935) and Australia. Meetings and conventions began to spring up in association with different chapters around the world, with major events taking place in Leeds in 1937 and New York (the first World Science Fiction Convention) in 1939.

Fans of science fiction came together to share ideas and debate opinions, listen to speeches from respected authors, and – beginning with the 1940 World SF Convention in Chicago – engage in mock weapon-fights and enjoy masquerade parties dressed up as their favourite science fiction characters. At the same time, as James (1994) maintains:

> Sf fans banded together not so much because they enjoyed reading sf, but rather because they had a particular sweeping vision of the place of humanity in the universe and of the tremendous potential which the future offered, which set them apart, they felt, from the humdrum world of the ordinary people around them. They were passionately committed to a technocratic approach to the world's problems. Like Gernsback, they thought that sf had an educational mission which was at the forefront of the progress of society towards a better world. Sf fans were the missionaries, and if at times they might refer to the infinitesimal brains of non-sf readers, at other times they certainly believed that it was their duty to proselytize and to win some of those brains over.
>
> (James 1994: 134–5)

Just as some journalists reporting on the conventions were quick to poke fun at the activities of the fans, certain 'highbrow' literary critics wasted little time in denouncing science fiction itself. Considering themselves to be arbiters of literary taste, in part due to their advocacy of so-called serious, respectable literature, they dismissed science fiction as being little more than whiz-bang escapist fantasy for adolescents. They frequently called into question its status as a genre, claiming that it lacked a sufficiently distinct history or tradition to fully qualify as a 'mainstream' category of literature. Moreover, the more vociferous voices among them insisted that science fiction offered the reader little by way of redeeming social value and, in any case, was contributing to the public's misunderstanding of scientific terms and principles.

Lines of criticism such as these and related ones articulated by critics outside of the communities of science fiction fans played an important role, paradoxically, in helping to sustain a collective sense of identity for those on the inside. Different accounts of the gradual evolution of science fiction into a more 'serious' or, for some, 'legitimate' literary genre will more often than not highlight the importance of John W. Campbell Jr's editorship of the pulp magazine *Astounding Stories*. Indeed, some go so far as to suggest that his period as editor, namely from 1937 to 1950, marks 'The Golden Age' of science fiction. Having been launched in January 1930 as *Astounding Stories of Super-Science* at a price of 20 cents per issue, the magazine had quickly gained a strong presence in the marketplace. Its first editorial, written by Campbell's predecessor Harry Bates, proclaimed: 'It is a magazine whose stories will anticipate the super-scientific achievements of To-morrow – whose stories will not only be strictly accurate in their science but will be vividly, dramatically and thrillingly told' (cited in Ackerman 1997: 125). The Campbell era of the magazine, which began in November 1937, is frequently described as the start of a Golden Age in part because, in the words of Ackerman (1997: 128), it soon provided 'a pyrotechnical cornucopia of the top sci-fi authors of all time and their glorious stories'. Campbell, a physics graduate from the Massachusetts Institute of Technology (MIT) and Duke University, introduced to his readership the work of writers such as Isaac Asimov, Leigh Brackett, Robert Heinlein, Raymond F. Jones, Henry Kuttner, C.L. Moore, Theodore Sturgeon, A.E. van Vogt, among many others. Stories by more established writers included those by Ray Bradbury, E.E. Smith, Jack Williamson and Campbell himself, under the pseudonym Don A. Stuart.

There was much to admire in the magazine's approach to science fiction, particularly for those readers anxious to move beyond the limitations previously associated with the Gernsbackian technology-oriented tradition of amazing devices and gadgets. For Campbell, in contrast, it was proposed that science fiction be best regarded as a literary medium aligned with the dictates of science itself. 'Scientific methodology', he argued, 'involves the proposition that a well-constructed theory will not only explain away known phenomena, but will also predict new and still undiscovered phenomena.' Science fiction, he added, 'tries to do much the same – and write up, in story form, what the results look like when applied not only to machines, but to human society as well' (cited in Clute and Nicholls 1999: 311). This shift in emphasis to better consider the social dimensions of technological change would have far-reaching implications for the concept of 'science' in many authors' visions of the future. While Campbell evidently saw in science 'the salvation, the raising of mankind', its representation was becoming more sensitive to the lived contradictions of 'scientific progress'. Awkward questions were being raised about how future worlds might be shaped by scientific innovations and inventions for better and for worse, thereby leading to stories which frequently challenged the cheery optimism characteristic of Gernsbackian fiction.

Also in 1937, the first British science fiction magazine, *Tales of Wonder*, appeared, and not before time. Prior to 1937, as James (1994: 42) points out, 'there had been no regular outlet for British sf in short form, and the unfettered speculation about the future which had become common in American sf remained largely absent: sf adventure and space exploration could find publication only as boys' fiction'. Not surprisingly, *Tales of Wonder* took its inspiration from its pulp counterparts in the US, frequently reprinting the work of writers from the other side of the Atlantic. Eventually joined by rival titles such as *New Worlds*, *Authentic Science Fiction*, and *Nebula Science Fiction*, these magazines presented writers such as Arthur C. Clarke, Eric Frank Russell and William F. Temple to wide acclaim. Elsewhere on the European continent, just as in Britain, science fiction was rapidly becoming an accepted genre label, one publishers were quick to use to advantage. In France, *Conquêtes* reprinted material from *Tales of Wonder* beginning in 1939, while the series 'Le Rayon fantastique' would prove to be even more successful. *Jules Verne-Magasinet* was hugely popular in Sweden, as was *Utopia-Magazin* in what was then West Germany, and *Galaxy* and 'I Romanzi di Urania' in Italy. In each case, these and related publications relied to a considerable extent on translations of US and, to a lesser extent, British science fiction writers, particularly following the end of the Second World War.

Science fiction's 'maturation' as a distinct genre was rapidly consolidated in the 1950s. For the pulp magazines, however, the end was in sight, due in no small part to the rise of paperback novels and television. Most titles had ceased publication by the middle of the decade, thereby taking with them public fears that their frequent misrepresentation of scientific concepts and principles was having a harmful effect on 'scientific literacy', especially among young people. In contrast, the sudden and dramatic rise of the paperback novel was providing new opportunities for science fiction to develop. Its commercial appeal was such that major publishers in countries like the US and Britain were moving quickly to recognize the genre as a distinctive, and profitable, category on their lists. At the same time, the term science fiction was being reinflected in a variety of other media contexts. Films such as *The War of the Worlds* (1953), *Spaceways* (1953), and *Forbidden Planet* (1956) were appearing alongside programmes on radio, such as *Journey into Space* (1954), as well as on the emergent medium of television: *The Quatermass Experiment* (1953), *Nineteen Eighty-Four* (1954) and *The Twilight Zone* (1959).

Across the media spectrum, it is difficult to overstate the significance of space flight in helping to create and sustain this boom in science fiction. The launch of Sputnik 1 by the then Soviet Union on 4 October 1957 ushered in an extraordinary wave of popular enthusiasm for the genre. 'To some of us steeped in the sf of the time,' writes Aldiss (1973: 245), 'space flight itself – and the launching of the first sputniks and satellites – seemed like an extension of sf, just as the Wellsian atomic bomb had done.' The subject of intense excitement in pulp stories, space flight 'had been the great dream, the great article of faith; suddenly it became hardware, and was involved in the politics of ordinary life' (Aldiss 1973: 245). It was also quickly embroiled in the politics of the Cold War, most strikingly following the US government's decision to counter the Soviet space programme's advances by launching a full-fledged 'space race'. On 31 January 1958, Explorer 1 became the first US satellite in space, and later that same year the National Aeronautics and Space Administration (NASA) was established. Elected to office in 1960, President John F. Kennedy moved quickly to approve the allocation of funds necessary to realize the objective of making US astronauts the first to land on the Moon. The possibilities of space exploration, so richly explored by writers of science fiction, were now being fiercely debated on the front pages of newspapers.

Alien worlds

The prospect of interplanetary travel, and with it the breaking free of earthly bonds, is arguably the most celebrated of the many themes which make up science fiction. A significant number of the precursors to the genre published in the seventeenth and eighteenth centuries – indeed it is often suggested the majority of such items – concern journeys to other worlds. Fascinating tales of 'cosmic voyages' include Johannes Kepler's *Somnium* or *The Dream* (1634), Francis Godwin's *The Man in the Moone* (1638) and Cyrano de Bergerac's *Histoire comique des états et empires de la lune* (or *Comical History of the States and Empires of the Moon*, 1657). Similarly considered to be noteworthy by historians is Gabriel Daniel's *A Voyage to the World of Cartesius* (1690), as well as Jonathan Swift's *Gulliver's Travels* (1726). In the case of the latter, it is the Voyage to Laputa in Book III which is frequently cited as having a particularly formative influence, not least on Voltaire's *Micromégas* (1752) with its hero, Sirius, who arrives on Earth in 1737 following visits to Saturn, Jupiter and Mars.

Stories of space flight have been catching the popular imagination ever since. Jules Verne's 1865 novel *De la terre à la lune* and its sequel *Autour de la lune* (published in English together as *From the Earth to the Moon* in 1873) usually earns acknowledgement as the first attempt to depict such flight in a scientifically realistic manner. The degree of realism here is relative, of course, given that the scientific principles that ostensibly make his space cannon feasible would in actuality easily crush the space travellers at the moment of takeoff. H.G. Wells would soon follow with *The First Men in the Moon* in 1901, a scientific romance criticized by Verne – as noted above – for failing to be sufficiently precise about the actual means of transport. Wells merely suggests that the spacecraft is propelled by a revolutionary anti-gravity metal called Cavorite, named after the scientist character, Cavor, whose invention of it is shrouded in mystery. With such technical details put to one side, attention focuses on Cavor and the tale's narrator, Bedford, as they encounter the insect-like Selenites living beneath the Moon's surface. The next year, 1902, would see aspects of both Verne and Wells's novels depicted in the first science fiction film epic, *Le Voyage dans la Lune*, by French cinema pioneer George Méliès. This film, at 21 minutes about four times as long as more typical films of the day, used highly innovative special effects to bring the narrative to life. In one of the more memorable comedic scenes, a cannon launches a rocket directly into the eye of the Moon, thereby causing it to grimace in pain. 'It is a movie of invention and good humour,' writes Shipman (1985: 14), 'but its success the world over amazed Méliès.'

It is hardly surprising that science fiction writers, together with their fans, were more excited than most that the dream of space flight might actually be realized in the 1960s. Events were unfolding at an extraordinary pace. In March 1961 President Kennedy decided to expedite the development of the Saturn rocket programme. On 12 April, Soviet cosmonaut Yuri Gagarin completed a circuit of the planet in a capsule called Vostok 1, making him the first human in space to survive the experience. On 25 May, in a televised speech before Congress, Kennedy formally set the goal of 'landing a man on the Moon and returning him safely to the Earth' before the decade was out. In the years to follow, both countries raced to achieve an astonishing array of 'firsts' in space flight, with the Soviet Union enjoying much more success. The Soviet Union was the first to send an unmanned probe to the moon and, in March 1965, cosmonaut Aleksei Leonov made the first EVA (extra vehicular activity) or spacewalk. In the US tragedy struck on 27 January 1967 when a fire at the launch complex 34 claimed the lives of three astronauts during a test of the Apollo spacecraft. Finally, on 20 July 1969, as hundreds of millions held their breath while listening to their radios or watching their television sets, Apollo 11's landing craft, Eagle, descended to the Moon. Astronauts Neil Armstrong (speaking the words: 'That's one small step for man, one giant leap for mankind') and Buzz Aldrin set about exploring the lunar surface, while Michael Collins orbited the Moon in the command module, called Columbia. All in all, six separate Apollo missions between 1969 and 1972 would carry twelve astronauts to the Moon. Over 300 hours would be spent on the surface, about 80 of which were put to use outside the landing crafts collecting rock samples, taking photographs and performing experiments to monitor the Moon's environment, among other activities. If the actual scientific dividend was minimal, the impact on 'America's self-esteem', its 'sense of place in the world', was profound.

'Science fiction come true' is reportedly how many people saw the landing of Apollo 11 on the Moon, although for some the symbolic victory of 'winning the space race' was a hollow one. The dream of space exploration, despite its recurrent expression in a language of freedom and democracy, was not one in which everyone was invited to participate as equals. Spigel (1997) is one of several critics to contend, for example, that 'the space race was predicated on racist and sexist barriers that effectively grounded "racially" marked Americans and women in general' (Spigel 1997: 47–8; see also Carter 1988; Watson 1990; McCurdy 1997). At the heart of this race, she argues, is a colonialist fantasy – what she terms 'white flight' – whereby Cold War strategies to 'contain' communism abroad interconnect with efforts to preserve racial segregation and female subordination in the household. Pointing to the ways groups excluded from space travel sought to challenge certain forms of racism, sexism and xenophobic nationalism underlying the 'giant leap for mankind', she maintains that the dream of the space race was also a dream to expand 'white suburbia and its middle-class, consumer-oriented family life into the reaches of outer space' (1997: 49).

Voices of dissent, while marginalized for the most part, were heard in the popular media. Of the examples Spigel cites, a passage from the September 1970 issue of *Ebony* magazine is particularly telling:

> Especially to the nation's black poor, watching on unpaid-for television sets in shacks and slums, the countdowns, the blastoffs, the orbitings and landings had the other-worldly alieness – though not the drama – of a science fiction movie. From Harlem to Watts, the first moon landing in July of last year was viewed cynically as one small step for 'The Man', and probably a giant step in the wrong direction for mankind.
>
> (cited in Spigel 1997: 64–5)

Opportunities for such criticisms, especially where the enormous financial investment in the space programme was set in relation to pressing issues such as inner-city poverty, were few and far between in the realm of popular entertainment. Instead, muted forms of criticism emerged in television programmes such as *Men Into Space* (1959–60). In one episode, 'First Woman on the Moon', female astronaut Renza Hale is not allowed to leave the rocket to explore the lunar landscape, the male astronauts insisting that she stay behind to attend to such concerns as anti-gravity cooking. Bored, she decides to venture out anyway, thereby sparking a panicked search for her by the others, leading one of them, her husband Joe, to exclaim: 'Your place is on earth at home where I know you're safe.' For Spigel, this and similar programmes such as *The Jetsons* (1962–3), *My Favorite Martian* (1963–6), *I Dream of Jeannie* (1965–70) and *It's About Time* (1966–7) merged the 1950s suburban family sitcom with science fiction fantasy, yet did so in a manner which introduced, in turn, new kinds of ideological tension. To varying degrees, she observes, the 'collision of [these] two unlikely forms presented viewers with the possibility of thinking about the social constraints of suburban life', often in potentially subversive ways (Spigel 1997: 58; see also Wolmark 1999).

Throughout the 1960s, the human exploration of space provided the backdrop for several science fiction series appearing on television. In Britain, what would become the world's longest-running science fiction programme, *Doctor Who*, was first broadcast on 23 November 1963. The opening episode appeared as the BBC returned its viewers to scheduled programmes following what had been 24 hours of almost non-stop coverage of the Kennedy assassination in Dallas, Texas. The topicality of space travel in the news had evidently generated the idea for a new series in the minds of BBC executives. 'Science fiction had become somewhat predictable,' recalls a historian of the programme, 'with rocket ships, spacemen and little green men from Mars' (Haining 1983: 16). The novelty of the Doctor, eventually revealed to be a Time Lord and self-proclaimed 'wanderer in the fourth dimension', revolved around his ability to travel back and forth in time and from one planet to the next via his time machine, the TARDIS (Time and Relative Dimensions in Space).

Resembling a police telephone box, the TARDIS operated – if somewhat unpredictably – to enable the Doctor and his fellow galactic travellers to experience adventures not dissimilar to the type found in pulp magazines. Despite the producer's initial desire to avoid 'bug-eyed monsters', episodes typically introduced an alien creature of some sort, most famously the Daleks in the first series. The Daleks, as Cook (1999) writes, became the Doctor's arch-enemies:

> the Daleks [are] horribly mutated creatures which have encased themselves completely in dome-shaped metal structures as protection from the radiation outside and which, in so doing, have turned themselves into the absolute personification of cold, machine-like evil: their complete disregard for other life-forms chillingly summed up by their famous, metallic war cry: 'Ex – term – in – ate!!.
>
> (Cook 1999: 117)

The extraordinary popularity of the Daleks with audiences meant the programme's educational agenda, with its promise 'never to transcend the scientific', was promptly displaced in the name of entertainment. Nevertheless, the Doctor himself, portrayed as a wise but eccentric inventor, was frequently made to represent the very embodiment of scientific endeavour. As Cook (1999) observes:

> When he pits himself against the machine-like Daleks, his is the old romantic image of the 'good scientist': the quirky, enquiring, free spirit committed to science as a means of individual emancipation and progress, versus the more modernist image of science gone wrong in the form of the coldly rational, dehumanising Daleks. In short: 'good' science as old-fashioned magic versus 'bad' science as new-fangled terror.
>
> (Cook 1999: 119)

Putting to one side the curious use of the word 'magic' in this context, it is apparent that *Doctor Who* succeeded in mobilizing competing visions of science and technology through its portrayal of alien figures like the Daleks. 'Compared to the dream of science as the progress of humanity embodied by the Doctor,' Cook (1999: 122) writes, the Daleks 'represent the corresponding nightmare vision of science as the road to utter dehumanisation'.

The importance of examining the representation of science and technology in *Doctor Who* is similarly taken up in Tulloch's contribution to

Tulloch and Jenkins's (1995) study *Science Fiction Audiences: Watching Doctor Who and Star Trek*. The remarkable success of the programme, as he points out, rested in part on its capacity to attract several different types of audiences, each with their own reading positions and subcultural affiliations. In seeking to explore these dynamics, Tulloch conducted a range of empirical audience studies with fans and followers of the programme, including one involving a focus group drawn from a 'para'-science fiction convention in Australia. Perhaps not surprisingly, given the members of this group's technical/technological or professional backgrounds, the discussion promptly turned to the special effects and gadgetry in the episode ('The Monster of Peladon') shown to them. By way of example:

> Phil: 'First of all, it was one of the better episodes from one of the better stories . . . There's more action, better effects, and you can't see that it's obviously faked. Everything there – provided you get the technology – could have worked.
>
> (cited in Tulloch and Jenkins 1995: 60)

This and related examples of their responses suggest to Tulloch that while the group members were positioned in the 'hard SF' camp, their interests were not strictly limited to the level of scientific 'gadgetry' and 'effects'.

Pointing to the diverse range of competencies the group brought to bear on their readings of the episode, Tulloch observes:

> Fan, generic, narrative and industry discourses were woven into a complex explanation which situated their discussion of action and effects. A concern for generically central concepts of SF (e.g. the balance between expansion and entropy, what it means to be alien, etc.) in fact led them to value *Doctor Who* well above the 'much more effects oriented' *Star Wars*. *Doctor Who* (though only when it was seen to match up to the possibilities of the genre) was 'more sophisticated than *Star Wars* will ever be'.
>
> (Tulloch and Jenkins 1995: 60)

Tulloch similarly notes that what some commentators have described as *Doctor Who*'s espousal of the 'ideological dominance of technological rationalism' is, for these fans at least, a source of real pleasure. The group's sense of the Doctor, for instance, was as a 'paternalistic' figure, that is, 'as a modern-day knight bringing the "new principles of physics and mechanics" to the post-medieval world' (1995: 60). Rather tellingly, several comments made by the group about the Doctor's use of technology (in the words of one group member, 'doing absolutely incredible things out of scrap metal') appeared to Tulloch to be indicative of what he calls 'the patriarchal ideology of science'. That is to say, this group's members, he observes, believe 'in a liberal and progressive social order based on scientific paternalism and technological curiosity' (1995: 61). This point leads him to add, in turn, that viewers like these ones 'are true descendants of science fiction's post-1920s' readership'.

By the time it came to a halt after 26 years, *Doctor Who* had become something of a national institution in the UK, and enjoyed a significant international following. The Doctor's ability to 'regenerate' his body meant that a succession of different actors (all white males) played the role, each contributing a different aspect to the character's evolution. Frequently associated with controversy, not least with respect to criticisms that it was too violent to be suitable for younger viewers, the programme mixed science with social commentary, often using the sharp edge of political satire to make its points. Evidently, according to one of the scriptwriters for the programme, 'no one in BBC management could be bothered to scrutinise the scripts for a "mere" science fiction show, enabling the writers to tackle themes forbidden to more respectable mainstream drama' (O'Brien 2000: 77).

If for some commentators *Doctor Who* was an exemplar of the darker, more pessimistic tendencies of British science fiction, few would dispute that *Star Trek* (1966–9) provided its viewers with a far brighter, reassuring vision of the future. Good ultimately prevails over evil in the *Star Trek* universe, a promise which is directly tied to its representation of science. Given the formative influence of the programme on science fiction more broadly, then, the next section's discussion will engage directly with *Star Trek* so as to pinpoint several pertinent issues.

The final frontier

Stretching across *Star Trek*'s episodes is a rudimentary system of 'alternative science' carefully invoked in order to lend credibility to what is a startling array of technological innovations. Arguably chief among these innovations is the capacity of the USS *Enterprise* to 'boldly go where no man [or woman] has gone before' at 'warp speed', that is, at a speed where travelling between star systems is plausible enough to be simply taken for granted from one planet's adventures to the next. The fact that the series was cancelled after only three seasons because of poor ratings, however, suggests that audiences in the 1960s might have had more difficulty with this concept than the programme's producers had envisaged. 'Star Trek probably came along too early,' maintained Gene Roddenberry, the programme's creator:

> Had man landed on the moon during our first or second year, the idea of space flight wouldn't have seemed so ludicrous to the mass audience. Star Trek probably would have stayed on the air. The eye of the world did not turn to space seriously as a future possibility until we were in our third year, and by then it was too late.
>
> (cited in Gross and Altman 1995: 79)

In any case, the 'technobabble' between crew members is presented as being indicative of 'twenty-third century science', and therefore must be seen to follow certain proscribed laws assiduously. The 'cultic' nature of *Star Trek* fandom is such, as Gregory (2000) argues, that it ensures that any anomaly is likely to be seized upon by viewers. This is especially so for those who, by dint of the seriousness of their engagement with the programme, describe themselves as Trekkers ('outsiders', in contrast, are more likely to refer to them disparagingly as 'Trekkies'). Star Trek's producers, all too conscious of the rigorous scrutiny to which each episode was being subjected, were in effect forced 'to create a scientific environment that is – in its own terms – consistent and credible, as well as ensuring a high degree of continuity between episodes' (Gregory 2000: 128).

'Science fiction like Star Trek', suggests physicist Stephen Hawking (1995: xi), 'is not only good fun but it also serves a serious purpose, that of expanding the human imagination.' In making the case that the physics that underlies *Star Trek* is particularly worthy of attention, he points to the importance of the programme's exploration of how 'the human spirit might respond to future developments in science' as well as its speculation 'on what those developments might be' (Hawking 1995: xii). What interests Hawking most is its representation of time travel (indeed, he made a guest appearance on *Star Trek: The Next Generation* to play a **virtual reality** game of poker with Data, Albert Einstein and Isaac Newton on the holodeck). Other scientists – such as Krauss (1995), a physics professor and author of the book *The Physics of Star Trek* – have focused on the scientific feasibility of phasers, tricorders, transporter beams, cloaking devices, deflector shields, matter-antimatter engines, and the like. Such enthusiasm for examining the extent to which the programme's representation of imaginary science and technology informs actual scientific practice is not shared by everyone, of course. Astronomer Carl Sagan (1997), for example, concedes that *Star Trek* has its charm, but promptly criticizes it for ignoring 'the most elementary scientific facts'. Turning his attention to Mr Spock, the *Enterprise*'s science officer, Sagan asserts: 'The idea that Mr Spock could be a cross between a human being and a life form independently evolved on the planet Vulcan is genetically far less probable than a successful cross of a man and an artichoke'

(Sagan 1997: 351). Even if one entertains the idea of aliens existing, he argues, they are likely to 'look devastatingly less human than Klingons and Romulans (and be at widely different levels of technology)', a view which leads him to conclude that 'Star Trek doesn't come to grips with evolution' (1997: 352; see also Mellor 2001b). Sagan's criticisms notwithstanding, the programme's success in engendering a sense of wonder about science appears to be undiminished among its still growing fan following.

Space as 'the final frontier' is an organizing thematic of *Star Trek*, with each episode elaborating upon Roddenberry's utopian vision of humankind's destiny. This destiny, from the vantage point of the twenty-third century, has seen the elimination of the conflict, poverty, inequality and prejudice characteristic of life on Earth in the twentieth century. The ascendancy of science, and with it the humanization of technology, is assured, while forms of discrimination such as sexism and racism have all but disappeared as 'illogical' aberrations in human behaviour. Not surprisingly, of course, such was hardly the case with respect to the production of the actual programme. Turning to the issue of sexism, McCurdy (1997) writes:

> When Gene Roddenberry produced the pilot [. . .] he promoted the topic of sexual equality by placing women second in command of the starship and dressing the crew in unisex uniforms. This proved too radical for the moguls at NBC [the television network], so after *Star Trek* began its weekly serialization, the woman returned as a nurse and the other female characters were given miniskirts to wear.
>
> (McCurdy 1997: 223)

Still, Roddenberry was successful in retaining one such female character, Lieutenant Urhura (played by Nichelle Nichols), on the starship's bridge as the communications officer. The significance of Nichols's presence, one of the first black actresses to become a network series regular, was not lost on audiences. Evidently, when Nichols considered leaving the series, she was contacted by civil rights leader Martin Luther King Jr, who persuaded her to stay so as to serve as an emblematic role model for black viewers. Nichols would later recall King as having said: 'Think about this, Nichelle. You have changed the face of television forever. They can never undo what they've done. The door is open!' (cited in Pounds 1999: 214). She later found herself at the centre of some public disquiet when in the third season episode, 'Plato's Stepchildren', her character was kissed by Captain Kirk (played by William Shatner). This was the first time a black woman and a white man had kissed in a US television series. Network officials, according to Gregory (2000), had been fearful that 'the scene might lead to a boycott of *Star Trek* in the southern states, despite the fact that the script clearly indicated that Kirk was being forced to kiss her by an alien being' (2000: 18). In Gregory's view, there is 'little doubt that such "controversies" contributed to the imminent cancellation of the series.'

Star Trek: The Next Generation (1987–94) placed a much greater emphasis on questions of human ethics and moral dilemmas than the original *Star Trek* series, a shift similarly evident in the types of scientific predicaments to be resolved. This enhanced concern with social issues, in many ways indicative of the 'new wave' movement from 'hard' to 'soft' science in science fiction writing more generally since the 1960s, created more diverse opportunities to represent the changing contexts of the human condition. 'Soft' science in this sense is understood in contrast with 'hard' in that it signalled a break from the latter's commitment to rigorously upholding the formal principles of 'real science' for their own sake. In other words, in sensitively exploring the moral quandaries science and technology pose for humanity, 'soft' science fiction encourages a more subtly nuanced, even contradictory appreciation of what it means to be 'human' in the first place. The fluidly contingent negotiation of subjective identities, and their embeddedness in the lived relations of class, gender, ethnicity, sexuality, and so forth, ensures that the nature of experience itself assumes a privileged status. In the case of *The Next Generation*'s android character Data (played by Brent Spiner), for example, his struggle to become fully human effectively blurs the otherwise rigid division between machine and sentient being. Data's longing to experience human emotions, often a source of comedic relief, is in turn sharply counterpoised by the ruthlessness of the Borg, evil enemies striving to 'assimilate' every species they encounter into their technological collective. The Borg represents the end of individuality as each conquered victim, including Captain Picard (played by Patrick Stewart) at one point, has implants melded into their body so as to ensure their absorption into the communal 'hive'. In the end, as one might expect, it is by exploiting the Borg's very denial of individual subjectivity that Picard and crew are able to save the day for humanity.

The tyranny of technological rationality, epitomized so chillingly by the Borg, is a dramatic theme taken up time and again in *Star Trek: Deep Space Nine* (1993–99) and *Star Trek: Voyager* (1995–2001). Some commentators discern here the influence of cyberpunk on 'mainstream' approaches to science fiction, arguing that public disillusionment, if not outright cynicism, about discourses of science and technology is making any attempt to portray them positively appear to be clichéd, or even anachronistic. Others point to the ways in which the masculinist underpinnings of much science fiction have been dramatically recast through the involvement of feminist writers, suggesting that these two *Star Trek* 'spin-offs' have become increasingly sensitive to gender and sexuality issues over time. *Voyager*, for example, which sees Captain Kathryn Janeway (played by Kate Mulgrew) at the helm, is far more character driven than the action-oriented original series, and *Deep Space Nine* even more so. Arguably discernible in both programmes are feminist critiques of male technophilia and scientism, as well as challenges to hierarchical configurations of ethnicity. It remains to be seen, at the time of writing, how such issues will be taken up in the latest *Star Trek* series, titled *Enterprise*. The new series is set in time about 100 years prior to the first, original *Star Trek*. Not only is the technology at a much more rudimentary stage of development – warp drive, for example, has just been invented – but also the social attitudes portrayed are less progressive than those represented in the series set further into the future.

Star Trek, in all of its different incarnations, continues to promote a social vision of the future where scientific rationality prevails in a law-governed universe. The optimism informing this vision, generally expressed within the terms of liberal humanism, stands in sharp contrast with the dystopias more typically found in contemporary science fiction. It also offers the viewer, in my reading, a sense of reassurance that scientific progress is both inevitable and desirable. In this way, *Star Trek*'s depiction of science finds its definition, in part, by its opposition to the larger 'postmodern turn' within the genre, where scientific concepts and principles are openly, and often playfully, transgressed. The postmodernist refusal to uphold the tenets of rationality renders problematic any appeal to 'value-neutral' science, just as it decisively undermines any attendant moral certitudes associated with a language of Truth. In the view of writers like Baudrillard (1994) at least, this dissolution leads to the effacement of the contradiction between the 'real' and the 'imaginary' itself. Here he makes the provocative charge that just as 'there is no real', it is also the case that 'the good old imaginary of science fiction is dead' (Baudrillard 1994: 121). Science fiction, he contends, is 'no longer anywhere, and it is everywhere', a claim that calls into question the very purpose of the genre category as we are all seen to be living a life of science fiction on an everyday basis.

Even if one hesitates to endorse Baudrillard's rather grand assertions, as I do, it is fair to say that it is increasingly apparent that precisely what counts as 'science fiction' is becoming evermore difficult to distinguish. Certainly *Star Trek*'s status as an exemplar of the genre can be challenged. For some, the programme is more aptly categorized as 'pulp sci-fi', with proper science *per se* providing little more than the occasional, and incidental, bit of 'technochat'. For others, including the physicist Krauss (1995: 174), the programme's emphasis on 'the potential role of science in the development of the human species' enables it to display 'the powerful connection between science and culture'. In any case, to even begin to grapple with the representation of science in *Star Trek* in all of its cultural complexity, one would need to look beyond the level of 'texts' (individual episodes, but also a cartoon series, several films and paperback novelizations) so as to also attend to the conditions of their production (see Whitfield and Roddenberry 1968; Solow and Justman 1996) and consumption (see Tulloch and Jenkins 1995; Harrison et al. 1996; Penley 1997).

Crucial to the latter dimension are media and industry responses to *Star Trek* (the very first episode broadcast on 8 September 1966 was panned by *Variety* as an 'incredible and dreary mess'; see Stark 1997: 259), as well as its extensive and highly lucrative merchandising. Generalized claims about the programme's popular reception need to recognize, of course, how its negotiation by different audience members is shaped by factors such as their respective age, class, gender, ethnicity and location (the original series, for example, was translated into almost 50 languages around the globe). In the case of *Star Trek* fandom, attention similarly needs to turn to 'official' and 'unauthorized' fan clubs, conventions, fan fiction and fanzines, Internet pages and chat-rooms, and even 'K/S' or 'slash' pornography (written by female fans who imagine Kirk and Spock as spacefaring lovers). This huge base of fans, as Penley (1997: 99) points out, 'represents one of the most important populist sites for debating issues of the human and the everyday relation to science and technology'.

Conclusion

This chapter has sought to show that one of the lasting achievements of science fiction has been to provide members of the public with an array of conceptual frameworks for engaging with scientific issues in ways that they find to be pleasurable and worthwhile. Science fiction speaks directly to people's concerns, fears, anxieties and desires, encouraging them to work through the possible implications of different scenarios while, at the same time, promising to keep them entertained in the process. For some people, science fiction will provide a more familiar context for these negotiations about the larger significance of a scientific concern (such as nuclear power, earthbound asteroids or human cloning) than, say, pertinent news items or documentary programmes. What might otherwise be regarded as a dauntingly complex issue, evidently requiring careful attention over time, can be creatively explored in a manner which makes sense to people in relation to their personal circumstances. Good science fiction has the capacity to rewrite our perceptions of science fact, and vice versa, in ways which can be as intellectually stimulating as they are readily accessible.

To imagine the future, as every reader of science fiction knows, is the first step in creating it. Still, at the same time, entertainment-driven portrayals of science recurrently fail the test of due scientific accuracy, as has been argued above. One need not accept the typical kinds of assumptions implied in generalized assertions about 'media effects' to nonetheless recognize that just as science fiction can be an important resource for people to draw upon, it can sometimes make an informed awareness of precisely what is at stake more difficult. New questions need to be asked, in my view, about the influence certain science fiction representations may be having for public discussion and deliberation about scientific developments and their consequences for society. In striving to investigate the varied ways in which media images of science circulate, critical inquiries clearly need to treat factual representations of science in conjunction with fictional ones if they are to adequately address the lived perceptions of ordinary people. It is to these debates concerning the media's role in the 'public understanding of science' that our attention turns in the next chapter.

Further reading

Disch, T.M. (1998) *The Dreams Our Stuff is Made of*. New York: Touchstone.

Krauss, L. (1995) *The Physics of Star Trek*. London: Flamingo.

James, E. (1994) *Science Fiction in the 20th Century*. Oxford: Oxford University Press.

Lambourne, R., Shallis, M. and Shortland, M. (1990) *Close Encounters? Science and Science Fiction*. Bristol: Adam Hilger.

McCurdy, H.E. (1997) *Space and the American Imagination*. Washington, DC: Smithsonian Institution Press.

Mitchell, W.J.T. (1998) *The Last Dinosaur Book*. Chicago: University of Chicago Press.

Penley, C. (1997) *NASA / TREK: Popular Science and Sex in America*. London: Verso.

Tulloch, J. and Jenkins, H. (1995) *Science Fiction Audiences*. London: Routledge.

Wolmark, J. (ed.) (1999) *Cybersexualities*. Edinburgh: Edinburgh University Press.

3 SCIENCE IN POPULAR CULTURE

> We have a society exquisitely dependent on science and technology, in which the average person understands hardly anything about science and technology. This is the clearest imaginable prescription for disaster – especially in a purported democracy. A fascination with pseudo-science is a dangerous foundation on which to base decisions about the environment, health care, defense and the many other urgent problems the nation and planet face.
>
> (Carl Sagan, astronomer)

Who among us will admit to believing that our planet is being visited by aliens from outer space, let alone claim to have actually seen an extraterrestrial spacecraft? How many of us know someone who is convinced that they have survived abduction from aliens? Not me, I hasten to add, but in saying so I nevertheless acknowledge that reportedly millions of people around the globe would answer questions like these ones in the affirmative.

Recent years have seen numerous public opinion polls conducted in different countries which appear to indicate that a startling number of individuals believe space aliens have visited Earth. One such Gallup poll in the US, for example, placed the figure at 27 per cent of Americans (Dean 1998: 3; see also Knight 2000). Many so-called 'ufologists' are adamant that the US government has known about the existence of unidentified flying objects (UFOs) for over 50 years, following what they insist was a crash landing of an alien flying saucer near the town of Roswell, New Mexico in July 1947. Rumours of the 'Roswell Incident' and subsequent 'alien autopsies' continue to flourish, due in large part to the efforts of certain ufologists, but also because of television programmes like *The X-Files* as well as films such as *Independence Day* (see also Campbell 2001). Scientific explanations for UFOs, to the extent that they are considered at all in the media, are routinely drowned out by allegations regarding covert government conspiracies and cover-ups. 'It's much more interesting to hear that there *is* extraterrestrial life,' as Sagan (1993: 4) observes, 'than to hear, well, we are just at the beginning of our studies and we don't yet know.'

Informing this chapter's exploration is a key question, namely what counts as 'science' in popular culture today? Debates over what can be properly said to constitute science are hardly new, of course, and no attempt will be made here to map their features in detail. Instead, the more pressing concern for our purposes is the growing concern among some scientists that the boundaries protecting scientific facts from pseudo-science speculation are crumbling – not least, it is argued, because of the irresponsible ways in which distinctions between the two are represented by the media. This chapter shall thus take as its point of departure the thesis that popular cultural images of science and scientists – appearing in both factual and fictional media contexts – are worthy of serious study. Indeed, it will be argued that these images are crucial factors which need to be carefully accounted for in any critical effort to rethink the central tenets of the 'public understanding of science' debate. It is precisely this issue of how to engage with such images, therefore, that this chapter aims to help recast afresh.

The truth is out there

A culture of paranoia pervades modern life. Sinister and inexplicable forces have been set loose upon us. The line between science and pseudo-science is hopelessly blurred, together with the distinction between the normal and the paranormal. Confusion reigns, chaos threatens, but the truth is out there for those brave enough to look for answers beyond the limitations of their own reality.

The tenor of these words, and certainly the phrase 'the truth is out there', will be instantly recognizable to fans of the popular television series *The X-Files* (Fox Broadcasting, USA). First appearing in September 1993, *The X-Files* features two FBI agents, Fox Mulder (played by David Duchovny) and Dana Scully (Gillian Anderson), who each week undertake to investigate cases 'out of the Bureau mainstream', that is, ones dealing with paranormal phenomena. Tensions between the two agents, which include the formulaic foibles of romantic attraction typical of primetime television, are largely framed by their sharply counterpoised approaches to solving mysteries. Mulder is a firm believer in the paranormal, his convictions having been decisively influenced by witnessing his sister's apparent abduction by aliens when he was 12 years old. His work in the FBI is dedicated to rescuing his sister, and to exposing what he regards as a government conspiracy to keep the evidence of interplanetary visitors from coming to light. Scully, a medical doctor with a 'background in the hard sciences', has been assigned to monitor Mulder's activities with an eye to 'debunking' or 'invalidating' his outlandish theories, a rather formidable task as it transpires. Expectations that a given episode's case will be neatly wrapped up with a clear-cut resolution are almost always undermined, as a number of possible explanations for the seemingly paranormal events being depicted are usually on offer. This absence of narrative closure is unsettling, even disturbing at times, yet frequently cited as one of the reasons for the programme's remarkable popularity with viewers.

The X-Files defies easy categorization in any one genre of television, as even its status as science fiction is questionable in light of its celebration of what some might regard as 'pseudo-science'. The basic tenets of modern science are repeatedly challenged, to say the least, and often in highly improbable ways. Introduced in the programme's pilot episode, for example, is the possibility of a UFO arriving to abduct people, a theme which quickly becomes a recurring element of *X-Files* narratives (and signalled in the opening credits by grainy images of a flying saucer). Its prominence throughout the series is unsurprising in the opinion of Lavery et al. (1996a), who argue that 'tales of alien abduction are unrivaled in contemporary America for their ability to combine the most terrifying aspects of paranormal experience with various cultural elements' (1996a: 7). These elements include, amongst others, 'science fiction; New Age obsessions with channeling, reincarnation, near-death experiences, and spiritual advancement; Byzantine government conspiracy stories, which include secret medical experiments upon unsuspecting citizens; and concerns with sexual abuse and **genetic engineering**' (1996a: 7). Interestingly, however, storylines revolving around such themes as alien abduction are consistently portrayed as being plausible without, at the same time, entirely ruling out alternative, scientifically based explanations. 'By always suggesting a possible human agency behind ostensible extraterrestrial encounters', Lavery et al. (1996a: 12) maintain, *X-Files* executive producer Chris Carter 'leaves all his options open and avoids sending anyone into ontological shock'. Unexplained events are typically linked to clandestine government operations, more often than not involving covert scientific activities being conducted so as to further certain conspiratorial objectives. The programme's representation of the paranormal thus readily merges with the fears of the paranoid ('Trust no one' being another *X-Files* catch-phrase). As Carter himself stated rather bluntly in an interview: 'You can lay on the bullshit really thick if you lay on a good scientific foundation. The show's only as scary as it is believable. Everything has to take place within the realm of extreme possibility' (cited in Lowry 1995: 33; see also Lavery et al. 1996b; Bellon 1999; Knight 2000; Campbell 2001; Howley 2001).

Some viewers of *The X-Files* have taken strong issue with its portrayal of science, not least certain members of the scientific community. Among the latter, as noted in Chapter 1 of this book, is British scientist Dawkins (1998). *The X-Files*, in his view, is anything but harmless fiction. As he writes:

> Imagine a television series in which two police officers solve a crime each week. Every week there is one black suspect and one white suspect. One of the two detectives is always biased towards the black suspect, the other biased towards the white. And, week after week, the black suspect turns out to have done it. So, what's wrong with that? After all, it's only fiction! Shocking as it is, I believe the analogy to be a completely fair one. I am not saying that supernaturalist propaganda is as dangerous or unpleasant as racist propaganda. But *The X-Files* systematically purveys an anti-rational view of the world which, by virtue of its recurrent persistence, is insidious.
>
> (Dawkins 1998: 28)

This line of criticism similarly informs Durant's (1998a) condemnation of the 'cult TV series', a programme he claims scientists loathe. Durant, who like Dawkins is a Professor of Public Understanding of Science, insists that unlike 'real science' *The X-Files* fails in its responsibility to make sense of the world. Scully and Mulder, for example, have their detective talk 'studded' with 'pseudo-scientific gibberish which would embarrass any self-respecting science-fiction writer', in his view. One reason scientists tend to hate the programme, he contends, is because in its popularity 'they see evidence of continuing public fondness for the oddball, the obscure and the occult'. Scientists, as 'custodians of genuine knowledge', to use his phrase, actively dislike the 'cheap imitation' of science being presented to viewers. That is to say, Durant believes many of them are offended 'by the parody of understanding that is willing to borrow scientific terminology (Scully and Mulder are forever talking about forces and fields and **DNA** and all the rest) as mere window-dressing for personal prejudice and superstition' (Durant 1998a: 5).

Still other commentators have pointed to Scully's increasingly precarious commitment to 'proper scientific analyses' (her role as defined in the pilot episode), contending that she is slowly coming around to accepting Mulder's rejection of scientific methods and principles. Some find deeply troubling the implied suggestion that there are limits to what can be understood by using scientific protocols, that there exist certain phenomena that openly defy rational explanation. Sagan (1997), for example, offers a pertinent assessment in this regard in his book *The Demon-Haunted World: Science as a Candle in the Dark*. There he proceeds to criticize the programme on the grounds that it 'pays lip-service to a sceptical examination of the paranormal', being 'skewed heavily towards the reality of alien abductions, strange powers and government complicity in covering up just about everything interesting' (Sagan 1997: 351). Sagan takes strong exception with the fact that only rarely does 'the paranormal claim turn out to be a hoax or a psychological aberration or a misunderstanding of the natural world', leading him to suggest that *The X-Files* would be providing a 'much greater public service' if such claims were 'systematically investigated' and in every case 'found to be explicable in prosaic terms' (1997: 351).

In response to Sagan's intervention, however, Penley (1997) contends that he has misread the programme. *The X-Files*, she argues, does not seek to 'substitute spiritualism for science', nor does it discourage a sceptical stance vis-à-vis the paranormal. 'Instead,' she writes, 'the show uses that which has been declared outside the bounds of science to challenge the categories and methods of science' (Penley 1997: 7–8). In terms of the kind of scepticism the programme promotes, she adds, it is a scepticism 'turned toward the complicity of the establishment science and the government in resisting people who believe that science is too important to be left to the scientists' (1997: 8). Moreover, Penley believes that Sagan's account misses the humour of *The X-Files*, especially the ways in which it 'mocks the pretensions and secretiveness of scientific-government institutions', nor does he seem to appreciate its use of irony. Many fans of the programme, as witnessed by Internet chat groups (online fans often call themselves 'X-Philes'), fanzines and the like, are thoughtfully attuned to its 'subtly intelligent fun with science', in her view. Some similarly find pleasure in its subversion of sexual stereotypes: 'The female Scully is rational, skeptical, and devoted to conventional scientific method, in contrast to her male partner's emotionality, attraction to the supernatural, and total lack of scientific superego' (1997: 8). At issue, from Penley's perspective, is the need to give members of the audience far more credit for being able to negotiate the complexities of the programme's representation of science than Sagan appears willing to concede. She insists that in being so 'doggedly vehement in his denunciations of popular culture', he was a 'humorless, antipopulist gatekeeper' whose criticisms need to be countered.

One avid fan of *The X-Files*, who also serves as one of its science consultants, is Simon (1999). In her book, *Monsters, Mutants and Missing Links: The Real Science behind The X-Files*, she describes how she became involved in the work of checking scripts for scientific inaccuracies prior to the production process. In addition to this responsibility, one of her key concerns is that Scully be consistently portrayed in a realistic manner. In Simon's view, Scully is the 'quintessential scientist' in that she 'gathers information and bases her hypotheses on that evidence. She also keeps partner Fox Mulder from rushing to unsupported conclusions [while] adhering to the scientifically sound precept that the simplest explanation is likely to be correct' (Simon 1999: 9). Still, as we have seen above, Scully is usually shown to be incorrect in her deductive reasoning, and at times questioning of scientific rationality itself. Therein lies the controversy, as Simon promptly acknowledges that the 'concept that science cannot explain all "unnatural" occurrences and the believable nature of the show's science-fiction scenarios leaves *The X-Files* open to critics who claim that it is "anti-science" ' (1999: 10).

Such criticism is unfair, in Simon's view, as were Scully to be correct more often, the programme would be far less compelling as science fiction. She suggests that those who see *The X-Files* as promoting pseudo-science are failing to recognize the extent to which scientists are being portrayed in a complimentary light. Notwithstanding the 'many bizarre and completely fictional creatures' that appear in different episodes, Simon insists that the scientific investigations of them have been based in reality. As she writes, the 'proper experiments are conducted; the correct microscopes are used; evidence is gathered and conclusions are based on that evidence' (1999: 10). In this way, she contends, the programme is far more grounded in 'real science' than is typical for primetime television, a virtue which is helping to make the world of science more attractive to young people. 'Critics who claim *The X-Files* is harmful to the public's awareness of science,' she maintains, 'would probably be amazed to learn how many students in my [first year] biology class point to the favourable portrayal of science and scientists on *The X-Files* as one reason for their interest in science' (1999: 250–1).

A related set of points are taken up in an editorial in the British journal *Nature* (27 August 1998). Here the argument is made that while belief in pseudo-science is a widespread problem in modern society, some detractors of *The X-Files* are failing to recognize that science itself 'can only progress darkly, up and down many blind alleys and false trails, from hypothesis to hypothesis'. Moreover, the programme is said to deserve credit for inviting its viewers to participate in making sense of the phenomena being depicted, 'rather than enforcing the exclusivity of a secular priesthood of which the public would be rightly suspicious'. Indeed, in questioning why *The X-Files* is so popular with audiences, the *Nature* editorial suggests that an answer may lie in the way it 'responds (or, if you hate it, panders) to the general fascination for the obscure and the inexplicable'. In any case, it is possible to interpret the programme's popularity as an indication that, in the words of the editorial, 'the public clearly has more of a feeling for the spirit of scientific enquiry than some give it credit for.' As a result, it follows, a 'rejection of the idea of [scientists as] custodians of truth does not reflect a preference for pseudo-science, but a dislike of being patronized'.

It is to this issue of how scientists are represented in media culture, not least as truth's ultimate custodians, that our attention turns in the next section.

Popular images of scientists

Scientists, it would seem, have a serious image problem. Research studies aiming to examine public attitudes to scientists typically make for grim reading, particularly where the opinions of young people are concerned. One such study, drawing upon an extensive survey of children in Leicester, England and Perth, Australia, found that children between 8 and 9 years of age typically see them as 'middle-aged white males who never have fun' (*BBC News On-Line*, 16 December 2000). In asking the children to draw a picture of a scientist, certain patterns became apparent. 'Boys never drew women and only very occasionally would a girl draw a female scientist,' observed one researcher, while it 'was also rare for a black or Asian student to draw a black or Asian scientist'. In an earlier study by researchers at the University of Bath, England, this time reporting on interviews with over 250 young people between the age of 15 and 17, it was similarly found that scientists were perceived in negative ways (*BBC News On-Line*, 21 December 1999). Specifically, scientists were recurrently regarded as being 'boring' and 'work obsessed', and stereotyped as being 'geeks' and 'dangerous cranks' who spend too much time in the laboratory. These kinds of research findings, not surprisingly, raise awkward questions about the teaching of science in schools, but also about the influence of images of scientists circulating in popular culture.

Here it is worth pausing to briefly trace the emergence of the word 'scientist' itself. Historians usually credit William Whewell, a professor of mineralogy and later of moral theology at Cambridge, with its introduction to English parlance. An active participant in early meetings of the British Association for the Advancement of Science in the 1830s, he is reported to have coined it on one such occasion. Whewell's two-volume book, *The Philosophy of the Inductive Sciences* (1840), similarly makes frequent references to 'scientist' as a more precise replacement for several related terms, most notably 'natural philosopher'. Significant here is his comment: 'we need very much a name to describe a cultivator of science in general. I should incline to call him a scientist' (cited in Williams 1983: 279). While crediting Whewell with formalizing this emphasis for the term, historians can nevertheless identify a wide range of earlier renderings of the concept it implies (one variant, 'sciencer', appeared in the sixteenth century). Gradually, from the mid-nineteenth century onwards, 'scientist' became firmly established in the public lexicon as a way to refer to practitioners of a specialized field of knowledge. The importance of this development for the larger professionalization of science would prove to be profound. 'It is one thing to think of the organized pursuit of knowledge that can be analysed independently of other social practices;' observes Fuller (1997: 35), 'it is quite another to think of its pursuit as a full-time job, a profession that requires specialist credentials' (see also Yeo 1993).

Returning to today, there can be little doubt that popular images of scientists – in the news and entertainment media alike – have a decisive impact upon public impressions of what scientists are like as people. 'Few outside the scientific world', Simon (1999: 10) notes, 'are personally acquainted with a scientist and therefore have only inaccurate fictional characterizations (usually unfavourable) on which to base their feelings.' Although Simon's observation overlooks the realm of factual media, it is fair to say that fictional images of the scientist would typically appear to have little to do with actual science and real-life scientists. While figures such as Isaac Newton, Marie Curie or Albert Einstein influenced the formation of such images, as Haynes (1994) contends they are notable exceptions to the general rule. Drawing upon her research into the status of the scientist as a cultural archetype, she maintains that it is fictional characters – among them Dr Faustus, Dr Frankenstein, Dr Moreau, Dr Jekyll, Dr Caligari and Dr Strangelove – that have proven to be far more consequential in shaping public perceptions. Not only do these fictional characters represent their respective creator's views about the role of science and technology in a particular social context, she argues, but also they have 'an additional historical significance both as ideological indicators of the changing perception of science over some seven centuries and as powerful images that give rise to new stereotypes' (Haynes 1994: 2).

In chronologically surveying the evolution of representations of scientists in Western literature from the late Middle Ages up to the present day, Haynes is interested in exploring how these fictional protagonists have contributed to what she describes as modern society's 'love-hate attitude towards science'. Interestingly, her examination of a diverse array of different materials, ranging from medieval images of alchemists to current depictions of **cyberpunks**, leads her to identify six recurrent stereotypes appearing across the texts under scrutiny (drawn mainly from literature, but also cinema). Briefly, she summarizes the features of these stereotypes as follows:

> *The alchemist*, who reappears at critical times as the obsessed or maniacal scientist. Driven to pursue an arcane intellectual goal that carries suggestions of ideological evil, this figure has been reincarnated recently as the sinister biologist producing new (and hence allegedly unlawful) species through the quasi-magical processes of genetic engineering.
>
> *The stupid virtuoso*, out of touch with the real world of social intercourse. This figure at first appears more comic than sinister, but he too comes with sinister implications. Preoccupied with the trivialities of his private world of science, he ignores his social responsibilities. His modern counterpart, the absent-minded professor of early twentieth-century films [. . .] is nevertheless an ineffectual figure, a moral failure by default.
>
> *The Romantic depiction of the unfeeling scientist* who has reneged on human relationships and suppressed all human affections in the cause of science. This has been the most enduring stereotype of all and still provides the most common image of the scientist in popular thinking, recurring repeatedly in twentieth-century plays, novels and films [. . .]
>
> *The heroic adventurer* in the physical or the intellectual world. Towering like a superman over his contemporaries, exploring new territories, or engaging with new concepts, this character emerges at periods of scientific optimism [. . .] More subtle analyses of such heroes, however, suggest the danger of their charismatic power as, in the guise of neo-imperialist space travellers, they impose their particular brand of colonization on the universe.
>
> *The helpless scientist*. This character has lost control either over his discovery (which, monsterlike, has grown beyond his expectations) or, as frequently happens in wartime, over the direction of its implementation. In recent decades this situation has been explored in relation to a whole panoply of environmental problems, of which scientists are frequently seen as the original perpetrators.
>
> *The scientist as idealist*. This figure represents the one unambiguously acceptable scientist, sometimes holding out the possibility of a scientifically sustained utopia with plenty and fulfillment for all but more frequently engaged in conflict with a technology-based system that fails to provide for individual human values.
>
> (Haynes 1994: 3–4, emphasis added)

Scientists, as quickly becomes apparent in considering these stereotypical images, are consistently being represented in negative terms. They are almost always men – female scientists, until fairly recently, being virtually unheard of in such imagery – who are ‘mad’ or ‘bad’. For each one striving to use science to make a positive contribution to improving public life, there are many more engaged in amoral activities with potentially dire implications for society. Significantly, Haynes (1994: 4) argues, 'these depictions have not only reflected writers' opinions of the science and scientists of their day; they have, in turn, provided a model for the contemporary evaluation of scientists and, by extension, of science itself'.

The recognition that these media depictions have a larger public significance beyond the realm of fictional entertainment suggests to Haynes (1994: 4) that we need to 'confront the widespread, often unacknowledged, fear of science and scientists in Western society'. This line of argument finds an echo in a range of studies concerned with popular images of scientists in a variety of different media contexts. In the realm of science fiction, for example, Hirsch (1962) examined the image of the scientist as presented in all science fiction stories published in the US during the period 1926 to 1950 (as listed in *Index to the Science Fiction Magazines, 1926–1950*). Findings from this content analysis, based on a sample of 300 stories, suggest that the proportion of scientists who are major characters declined steadily over this period. The relative number of scientists depicted as 'heroes' similarly dropped, while the likelihood of them appearing as 'villains' increased (followed, in turn, by 'businessmen' as the second most frequent villains). The portrayed status of scientists in imaginary societies suggests that they tend to be represented as the 'legitimate' elite in comparison with non-scientists, although by the later phases 'the naïve adulation of the omnipotent and omniscient scientist is no longer a feature of the genre as it had been in its childhood' (Hirsch 1962: 265). These later stories also tend to picture scientists in 'bureaucratic' settings, enmeshed in a range of interpersonal and institutional pressures, as opposed to the 'independent' or 'gentleman scientist' more typical of the earlier stories. Overall, the most frequent type of narrative conflict revolves around interpersonal relations (usually involving a 'love interest'), followed by the effects of technology (a 'Frankenstein' theme is recurrent), international conflict and, lastly, interplanetary conflict. The majority of the stories, as one might expect, have 'happy' endings. By the end of the period under scrutiny, according to Hirsch (1962: 267), scientists 'are no longer either supermen or stereotyped villains but real human beings who are facing moral dilemmas and who recognize that science alone is an inadequate guide for the choices they must make' (see also Parrinder 1990).

LaFollette (1990), in her study of images of science in popular magazines published in the US between 1910 and 1955, highlights the recurrently sexist ways in which female scientists were depicted. It was typically the case in titles like *The Saturday Evening Post*, *American Magazine* or *Scribner's* appearing over this period, she argues, that the so-called 'average scientist' was mythologized as being 'male, white, brilliant, energetic, rational and dispassionate' (LaFollette 1990: 6). Success in science, according to most of these magazine descriptions, required certain 'masculine' attributes. LaFollette shows that these attributes, including intellectual objectivity, physical strength and emotional detachment, worked to define scientific research in such a way that the participation of women was judged to be socially inappropriate – if not outright impossible for them to accomplish in any case. A woman wanting to be scientist, it followed, must be an 'abnormal' person, someone acting contrary to nature.

In depicting women as 'minority characters' in the 'drama of science', these magazines usually aligned them in one of two extreme positions. That is to say, they typically appeared either as subordinate assistants to their male counterparts, or as 'super-scientists'. In the case of the latter position, a tiny number of exceptional female scientists – such as Marie Curie or Margaret Mead, for example – became glamorous stars on the pages of some of these magazines. As LaFollette writes:

> To be a woman scientist was not only to be different from other people and from other women – mentally, physically, perhaps even sexually – but also to be strong enough to perform like a star and to do so alone. The male scientist's support structure – his wife, children, assistants, colleagues, neighbors – were ready at his side; the female scientist either had to be a Joan of Arc, alone at the top, or a dual-role housewife/scientist, one hand holding the mop (or child), the other steering the telescope toward the stars.
>
> (LaFollette 1990: 93–4)

This focus on 'superwomen' scientists, whether as audacious explorers, dedicated saints or the like, meant in turn that women were seldom made the subject of articles about 'normal', 'ordinary' or 'everyday' scientists. Consequently, by 'implying that the options of a regular family life, home and possibly motherhood were closed for all but the superwomen,' LaFollette (1990: 94) contends, 'this public image may have dissuaded many students from choosing scientific careers' (see also Broks 1996).

Several researchers have turned their attention to the images of scientists presented in mainstream feature films. In addition to Haynes's (1994) research mentioned above, Elena (1997), for example, examines the film *Madame Curie*, produced by MGM in 1943, citing its formative influence on cinematic treatments of female scientists to follow to this day. 'The film's viewpoint', he argues, 'is consistently one of surprise at the phenomenon of a woman scientist' (Elena 1997: 275). Evidently one of Hollywood's first 'biopics' of scientists, *Madame Curie* depicts Marie Curie, the discoverer of radium, as little more than a 'research assistant who is permanently subordinate to a male scientist', in this case Pierre Curie. Jones (1997) surveys several post-war British films appearing over the period 1945 to 1970 in order to pinpoint the most prominent stereotypical representations of scientists. The three main stereotypes identified are the artists, who are unworldly innocents; the destroyers, namely scientists who in the course of their work (intentionally or not) produce a result that harms themselves or others; and the Boffins, that is, scientists in the employ of the government, usually involved in weapons production (Jones 1997: 34). It is the last of these stereotypes which most intrigues Jones, who argues that such figures are usually presented in a positive light due to their efforts to use science to help 'our' side in the war. A key film here is *The Dam Busters* (1954), where engineer Barnes Wallis, who invented the bouncing bomb, is memorably portrayed by actor Michael Redgrave as the archetypal Boffin. In the case of US cinema, Jones suggests that the best known example of a Boffin is the title character in *Dr Strangelove* (1964), played by Peter Sellers. In his role as 'Director of Weapons Research and Development' for the US government, Strangelove assumes an independent stance from the politicians and military officials at the war council – a character reportedly developed with Edward Teller, the so-called 'father of the H-bomb', in mind.

Turning to the newspaper press, Nelkin's (1995) analysis of US journalism suggests that scientists 'appear to be remote but superior wizards, culturally isolated from the mainstream of society' (Nelkin 1995: 14). This kind of heroic imagery, she suggests, is most prominent in newspaper reports of prestigious scientists, especially where Nobel laureates are the story. In examining press items concerning the annual announcement of **Nobel Prize** winners, she identifies several characteristic features of the coverage that have appeared with 'stunning regularity' over recent decades. Stories tend to focus, for example, on the national affiliations and 'stellar qualities' of the Nobelists, typically following the stylistic conventions of either sports writing or reports of the Academy Awards or 'Oscars'. Scientists from different nations are described as rivals pitted against one another in pursuit of honour and glory. Customarily obscured from view is the international cooperation routinely associated with scientific endeavours, as well as the actual details of the prizewinner's research or its scientific significance. 'Rather,' Nelkin (1995: 16) writes, 'when the research is described, it appears as an esoteric, mysterious activity that is beyond the comprehension of normal human beings'. Similarly discernible is a gender politic whereby the successful male scientists are presented as being 'above the mundane world and totally absorbed in their work', while female laureates, in marked contrast, are to be 'admired for fitting in and for balancing domestic with professional activities' (1995: 19). The overwhelming message of popular press accounts, Nelkin argues, is that 'the successful woman scientist must have the ability to do everything – to be feminine, motherly, and to achieve as well' (see also Shachar 2000).

In the realm of young people's media, the portrayal of scientists in children's educational science programmes has been the subject of several research studies. Steinke and Long (1996), for example, analyse images of female scientists on four popular science television series, namely *Mr Wizard's World*, *Beakman's World*, *Bill Nye the Science Guy* and *Newton's Apple*, broadcast in the USA. Examining five episodes from each series, the authors sought to determine 'whether female scientists were represented equally on these programs in regard to the frequency of appearance and professional status' (Steinke and Long 1996: 98). Not surprisingly, some of the programmes faired better than others, but overall the study's findings indicate that while boys and girls were equally represented, male scientists appeared twice as often as did female scientists. Moreover, Steinke and Long (1996: 106) argue, the 'majority of female characters shown on these programs were cast in secondary roles of lower prestige in the scientific community', that is, as either students or laboratory assistants. Accordingly, while in the authors' view all of the programmes were valuable sources of informal science learning opportunities for children, much more needs to be done to ensure that female scientists are consistently portrayed in a realistic manner. 'Images of scientists on children's educational science programs', Steinke and Long (1996: 110) maintain, 'should show young viewers that it is ability, not gender, that matters – and that it is neither unusual nor atypical to see a female scientist running a lab of her own.' Encouragingly, a follow-up study of three of these four programmes along with *Magic School Bus* notes some recent improvements in the depiction of female scientists. However, it points out that in looking at the ethnicity of the different characters across the range of programmes, 'minorities were significantly less likely to be labeled as scientists, minority scientists spent much less time on screen than did Caucasian scientists, and there were significantly fewer minority characters than Caucasian characters' (Long et al. 2001: 255). Young viewers, the study concludes, are being encouraged to believe that 'science seems to belong to Caucasians' (2001: 265; see also McArthur 1998).

What, then, do scientists themselves make of their media image? Somewhat paradoxically, research suggests that there are scientists who actually attempt to conform to outdated stereotypes when they speak or write for 'lay people'. Shortland and Gregory (1991), in a handbook aimed at helping scientists to communicate with the public more effectively, contend that some scientists feel the need to present themselves as a 'serious, pulseless automaton subjecting Nature to the unassailable scientific method' (1991: 10). While these scientists recognize that this kind of stereotype is ridiculous, many strive to emulate it nonetheless. As a result, according to Shortland and Gregory, they

> lose their personality and enthusiasm for science, and spout facts with all the subtlety of an exploding encyclopaedia. Some people hide behind the stereotype, and use it as an excuse for keeping themselves to themselves. Other people think that showing their feelings is 'acting' or 'pretending', and that pretence has no place in science, which thrives on the honesty and integrity of scientists.
>
> (Shortland and Gregory 1991: 10)

It is at this level, Shortland and Gregory maintain, that such stereotypical images of the scientist can do considerable harm. Indeed, in their view, images of scientists bereft of emotions, temperament and feelings contribute to the creation of barriers between the scientific community and the general public. Speaking directly to scientists, they advise: 'The more effort you make to dislodge that image by portraying the excitement, challenges, and human qualities of science, the more likely it is that ordinary people will be interested in what you have to say' (Shortland and Gregory 1991: 22; see also Myers 1990).

Implied in this line of argument, rather tellingly, are a series of assumptions about how the 'scientific community' can and should relate to 'the general public' and vice versa. New questions promptly arise as to who defines 'what counts' as science, how science can be made more 'interesting' or 'popular' with members of the public, and why these dynamics are seen to be so consequential for society at large. Many of the (at times acrimonious) debates around these issues, as will be shown in the next section, are organised into an area of inquiry broadly encompassed under the rubric the 'public understanding of science'.

Public understandings of science

At first glance, the phrase the 'public understanding of science' appears harmless enough, possibly making one wonder how it ever became the site of such controversy. Who, it might be tempting to ask, is actually opposed to the public understanding of science? Surely everyone is in favour of the idea that members of the public should take an interest in science, even if they might disagree about how to go about achieving it? Where things get much more interesting, of course, is at the point where different users of the phrase slow down to specify what they mean by 'the public', 'understanding' and 'science'. More often than not, the ensuing debates about how best to define these terms turn into rather lively events, with each participant keen to press their preferred formulations to advantage. Disputes over precisely what is at stake in both conceptual and methodological terms have been raging for some time now, and are showing little sign that they are about to subside or be resolved once and for all.

The public understanding of science, as Durant (1992a: 7) suggests, is something of a catch-phrase that 'has come to stand for concern about the relationship between science and technology, on the one hand, and the general public, on the other'. In Britain, this concern was addressed by a Royal Society of London working group in the early 1980s, leading to the publication of a highly influential report in 1985, titled *The Public Understanding of Science* (Bodmer 1985). This report – often referred to as the Bodmer Report after the group's chairperson, geneticist Sir Walter Bodmer – provided a range of recommendations concerning how the public understanding of science might be improved. In Durant's words, the report 'lamented what it took to be the inadequately low level of scientific understanding among the public, and it emphasized the need for the scientific, the industrial and the educational communities to be more active in overcoming public ignorance' (Durant 1992b: 3; see also Pearson et al. 1997). One of its key efforts in this regard was to persuade the Economic and Social Research Council (ESRC) to fund a major programme of research in this area. A further effort was the establishment in 1986 of a Committee on the Public Understanding of Science (COPUS) as a joint venture of the Royal Society, the Royal Institution (founded in 1799 with the remit to promote 'the application of Science to the common Purposes of Life') and the British Association for the Advancement of Science.

The principal aim set for COPUS was to encourage a greater public engagement with science. Accordingly, a series of promotional activities intended to further enhance public awareness and understanding of science were launched, a key objective being to contribute to the improvement of standards of science communication in the research community, schools, media, museums, publishing and the civil service. More specifically, COPUS promptly set about developing several interrelated initiatives, such as making small grants available for pertinent research, hosting media training workshops for scientists, funding community-based projects, and awarding (with the Science Museum) a major prize for the best popular science book. It also claimed as an early success the effort to have 'a character who happens to be a scientist written into a TV soap opera', although it admitted that there was 'no indication that any TV company is ready to pick up the idea of a science soap' (Bodmer and Wilkins 1992: 8).

On a governmental level, COPUS has sought to intervene in the processes of public policy formation for science issues, including those associated with Green Papers (e.g. on the BBC in 1993) and science-related White Papers (e.g. *Realising Our Potential* also in 1993), as well as various parliamentary committees on science and technology. Recent efforts to improve public access to science, according to promotional materials available from its information office, include: a broadcast fund in collaboration with the Educational Broadcasting Services Trust, a fund-raising mechanism for science and technology on television and radio, and a collaboration with the Museums and Galleries Commission to encourage non-science museums to include science and technology activities. At the same time, a number of new projects are being introduced, including one involving pairings between Royal Society research fellows and 'science neutral' Members of Parliament (MPs) – the hope being, evidently, that this pairing arrangement will offer both parties an insight into each other's role and priorities. This and related projects are informing the current reorganization of COPUS in a way which promises to extend its national role as a resource for information about science communication.

Many of the efforts currently being mobilized in Britain under the broad rubric of the public understanding of science are rapidly converging into an international movement. Pertinent rallying points for its advocates include 'la culture scientifique' in France, for example, or 'scientific literacy' in countries such as the US. Important in the case of the latter is the Scope, Sequence and Coordination (SS&C) Project of the National Science Teachers Association, as well as Project 2061, the long-term initiative of the American Association for the Advancement of Science (AAAS). Taking its name from the anticipated date when Halley's comet will next visit, Project 2061 aims to reform the educational system across the country so as to ensure that all high-school graduates will be sufficiently 'science literate' so as to be able to deal with related social issues. This emphasis on 'science literacy' finds an echo in the activities of a wide range of groups around the world, such as in South Africa where a joint venture with Project 2061 is underway with that country's Department of Arts, Culture, Science and Technology. Meanwhile Euroscience, with its members from across 38 European countries, provides an 'open forum' for scientists, industrialists and citizens interested in science and technology, and seeks to help shape science and technology policies. In China, the Chinese Association for Science and Technology is introducing initiatives to improve science education, its own public opinion research suggesting that less than 2 per cent of its population can demonstrate 'basic scientific literacy'. In

Malaysia, support for public awareness of science and technology is being co-ordinated by the Malaysian Science and Technology Information Centre. In Australia, as a final example, the Federation of Australian Scientific and Technological Societies (FASTS) represents the interests of some 60,000 scientists and technologists in that country. Operating as a pressure group of sorts, it intervenes in public debates so as to influence the formulation of science and technology policy, especially where science education is concerned.

Looking across national contexts, Henriksen and Frøyland (2000) usefully outline four of the key arguments that are often presented to justify pursuing scientific literacy in the population:

- *The practical argument*: people need an understanding of science and (even more) technology to handle everyday life in a science- and technology-dominated society.
- *The democratic (civic) argument*: people need an understanding of science to relate to the many complex science-related issues that confront citizens of modern democracies.
- *The cultural argument*: science is part of our cultural heritage and has profoundly influenced our view of the world and of humankind's place in it; thus, one needs a grasp of what science is in order to understand culture. Moreover, knowing something about the objects and phenomena in the world that surrounds us is a source of joy and fulfillment for the individual.
- *The economic (professional) argument*: a scientifically literate workforce is necessary for a sound and flourishing economy in most countries.

(Henriksen and Frøyland 2000: 393)

As becomes clear in considering these and related points, proposals to improve 'science literacy' sound laudable enough – who, for example, would wish to make the case for 'science illiteracy'? Still, precisely what is meant by 'science literacy', and the extent to which it is actually tenable as an ideal worth striving for, is anything but straightforward. Normative claims about what members of the public 'ought to know about science' will be judged according to rather different criteria depending on who is doing the defining, and what interests are at stake (see Durant 1993; Irwin 1995; POST (Parliamentary Office of Science and Technology) 1995; Mellor 2001b; Nowotony et al. 2001).

Underpinning much of the rhetoric associated with this kind of 'science literacy' agenda – to varying degrees, of course, depending on the specific project under scrutiny – is what Ziman (1992, 2000) aptly describes as a 'deficiency' model of science and its social role. In his words:

> According to most scientists, the great majority of ordinary people have very little understanding of science. The outstanding feature of the situation is deemed to be public *ignorance*. The primary question is, 'What do people not know – and for goodness' sake, why not?' The problem is perceived as a *deficiency*, which must by all means be overcome. The basic measure of progress in public understanding of science is taken to be how much more science people can be made to understand.
>
> (Ziman 1992: 13–14)

For Ziman, himself a theoretical physicist by training, there is nothing new in this apparent 'knowledge gap' between the world of science and the world at large. As he acknowledges, efforts to first narrow and then close this gap have been made by a number of science associations in different countries over the past two centuries. The fact that these efforts do not appear to have had much effect, however, suggests to Ziman that the deficiency model is problematic in a number of ways. Chief amongst its difficulties, he contends, 'is that "science" is not a well-bounded, coherent entity, capable of being more or less "understood" ' (Ziman 1992: 15). That is to say, precisely 'what counts as science is defined very differently by different people – or even by the same people under different circumstances' (1992: 15–16). Scientists themselves, he adds, do not uphold a clear and consistent notion of what 'science' covers, leading him to argue that science 'is not a special type of knowledge that only starts to be misrepresented and misunderstood outside well-defined boundaries by people who simply do not know any better' (1992: 16).

Accordingly, familiar assertions about the alleged lack of scientific literacy in the public, and with it certain corresponding assumptions about how science needs to be 'popularized' to overcome this public 'ignorance', continue to be hotly contested. For some critics of the public understanding of science movement, this kind of language is simply inappropriate. From a feminist perspective, as Barr and Birke (1998: 15) argue, the implied deficit model 'is likely to be gendered (who is most likely to be seen as deficient in knowledge?) and to ignore the wider context of knowing' (see also McNeil 1987; McNeil and Franklin 1991). This emphasis on the multiplicity of different ways of knowing otherwise obscured by the deficit model is similarly addressed by Gregory and Miller (2001). 'Within this model,' they write, 'scientists are the providers of all knowledge, and the arbiters of just what should be provided, to an empty-headed public' (2001: 61). It follows from there that there are several attractions of this model for scientists: 'it places them at the top of the epistemic heap; it leads to a clear program of action – "Fill those heads!"; and, the results of the program should be empirically testable in surveys of public understanding of science' (Gregory and Miller 2001: 61–2). However, where this approach 'fails woefully', according to Gregory and Miller, is when the science in question is under dispute within the scientific community, or when it is still under development (science-in-the-making). Therein lies the problem, they argue, as 'much of the science the public needs to know about is either hotly contested or is still on the assembly line' (2001: 62; see also Labinger and Collins 2001).

Efforts to engage with how members of the public actually *use* science should ensure, in my view, that any top-down, zero-sum formulation of 'understanding' versus 'ignorance' will collapse under its own weight. Pertinent in this context is Bucchi's (1998) unpacking of the kinds of precepts typically underlying public discourses about 'science and the media' or 'communication of science to the public', work that leads him to argue that they rest upon an untenable set of truisms. That is to say, in effect these discourses revolve around the same truisms found in what he characterizes as the 'canonical account' of the communicative relationships between science and society (see also Best and Kellner 2001). This account is 'canonical', Bucchi maintains, to the extent that both scientists and journalists alike commonly uphold it, and in so doing make it the dominant view of the popularization process. His summary of these basic truisms is worth citing at length here:

1 The scientific enterprise has become too specialized and complicated to be understood by the general public. For example, in December 1919, when two solar eclipses had finally confirmed Einstein's general theory of relativity, *The New York Times* gave great emphasis to a comment allegedly made by Einstein himself: 'In the world there are no more than a dozen persons who could understand my theory.'
2 Therefore, a form of mediation is needed in order to make scientific achievements more suitable and accessible for the public. This mediation requires the intervention of a new professional figure: a 'third person' (in general, the science journalist) who can manage to bridge the gap between scientists and the non-scientific audience, by understanding the former and communicating their ideas to the latter. The introduction of this 'third person' is essential for [scientific] researchers as they can claim to be completely indifferent and extraneous to the process. It is the journalist, after all, who takes their theories and sentences and rearranges them for the public where 're-arranging', according to the scientists, usually means 'distorting'. For example, 'Natural knowledge . . . is perceived as watered down and then trickled down for popular consumption, along the way losing theoretical content.' [see Cooter and Pumfrey 1994: 248]

> 3 This mediation is most often described through the metaphor of linguistic translation. As a sort of interpreter, the 'third person' [e.g. science journalist] should simply accomplish the task of reformulating scientific discourse in more simple words. From this point of view, the problem of communicating science to the public, then, is reduced to a mere matter of linguistic competence.
>
> (Bucchi 1998: 3)

It quickly becomes apparent that according to this 'canonical account' there exists a sharp division between scientific discourse, on the one hand, and the public discourse of science, on the other. The task of the scientist is thus one of producing 'pure' knowledge so that it then may be offered to interested 'non-experts' in a simplified form (Bucchi 1998: 3–4).

It is hardly surprising that Bucchi wastes little time in challenging this set of apparent truisms. The canonical account, he points out, represents an idealized vision of the public communication of science, where the main groups involved in the process – scientists, journalists and the public – perform mutually exclusive roles. This vision, he argues, has its roots in the professional ideologies of scientists and journalists, and continues to be so influential largely because it serves useful purposes. For scientists, the account enables them to deny any direct involvement in the so-called popularization process, thereby leaving them 'free to deprecate its faults and excesses, namely inaccuracy and spectacularization' (1998: 4). Journalists, in contrast, 'need it in order to justify their role and to give sharper focus to the nature of their task' (1998: 4). Such a dynamic between these two groups is understandably fraught with tensions at a number of different levels. In striving to prepare a fair and balanced report about a scientific finding or development, the problem of how to best negotiate a range of contending pressures comes to the fore. The very conventions of journalistic practice, like the institutional constraints of the news organization, impose severe limits on what can be reported and how – to say nothing of the reporter's need to try to accommodate the interests of the scientists involved.

These kinds of tensions engender, in turn, what Bucchi describes as a 'blame the messenger' perspective. That is to say, wherever a comparison is drawn between 'original' scientific ideas and their portrayal in the media, the media coverage is almost always found to be wanting. It follows that the resultant 'misrepresentation' or 'distortion' of scientific ideas in the media inevitably results in widespread confusion and anxiety, thereby leading to, amongst other concerns, an 'insufficient appreciation of scientific achievement by the public itself' (Bucchi 1998: 4). By this logic, then, the popularization process is said to consist of a 'unidirectional, linear communication transfer from one sender (the scientific community) to a completely passive receiver (the broad, uninformed public)', a process that 'should in no way affect the nature and content of original information' (1998: 5). Thus basic to this canonical account, as Bucchi (1998: 5) observes, is the assumption that 'lay audiences simply absorb, in an impoverished and lessened form, ideas which stem from scientific activity.' Indeed, as he proceeds to add, given 'the proper transmission of information, people will be led along the "royal avenue" to scientific awareness' (1998: 5). It is up to the media, then, to ensure that scientific issues are 'disseminated' in a manner which satisfies scientists' assessments of accuracy and reception levels.

In calling into question the canonical account of the popularization process, researchers like Bucchi are seeking to reintroduce into its underlying assumptions a degree of complexity that appears to have been otherwise denied for the sake of conceptual unity. Not only does this account appeal to a normative vision of how popular knowledge about science circulates across society, but also it prefigures a totalized, unidirectional understanding of how such knowledge is reappropriated (to varying extents) by members of the public in the context of their everyday lives. Moreover, critics like Goldsmith (1986) point to the specialist fragmentation of science, as well as the rapid speed of its development, to argue that it is not simply a matter of somehow making scientific views clear through popularization. Science, he observes, tends to be presented by science writers as a 'collection of facts', and their explanations of these facts are usually incomprehensible to the non-scientist. In his words:

> Popularization of science as a social undertaking designed for a privileged minority has no place in a democratic society and rapidly changing society. We must rid ourselves of this concept of 'popularization', which is derived from Victorian, condescending, do-goodism, from a patronizing aloofness irrelevant to today's needs. We do not need to make science 'popular' in a world in which science is of key importance [. . .] The catastrophic power of the applications of modern science confronts each one of us with profound moral issues. Popularization of science has tended to alienate people from science. What we need is more public understanding of science, and more public appreciation of the impact of science.
>
> (Goldsmith 1986: 14)

This and related modes of critique have succeeded in rendering problematic several of the tacit presuppositions underpinning the popularization concept. Several efforts to recast the tenets of the canonical account have begun by undertaking a careful dissolution of the rigid dichotomy it imposes between scientific discourse and the public communication of science. In recognizing that this type of dichotomy can be sustained only at the level of an abstract conceptual model, and even then with difficulty, attention has turned to investigate the complex ways in which these dynamics actually work in given situational contexts.

This emphasis on exploring how scientific knowledge is embedded in diverse situational contexts has enabled critics of the 'deficiency' or 'deficit' model to push research into new, exciting directions. Particularly significant, in my view, are the contributions of those researchers seeking to challenge what they perceive to be the dominant public understanding of science agenda by bringing attention to bear on the broad political culture of science. Much of this work assumes a critical stance vis-à-vis the central tenets of the public understanding of science movement. In the case of Wynne's (1992) research in a British context, for example, 'the re-emergence of the public understanding of science issue in the mid-1980s can be seen as part of the scientific establishment's anxious response to a legitimation vacuum which threatened the well-being and social standing of science' (Wynne 1992: 38). It is this legitimation vacuum, it follows, that underlies the fears expressed about public 'ignorance' of science. This situation is somewhat ironic, Wynne observes, given that this vacuum is the product of the way science has been effectively distanced from the ordinary public in the past, in part so as to better justify its claim to public funds. Science, he argues, 'now finds itself hoist with its very own petard, namely the cultural alienation whose establishment it actively, if innocently, promoted' (1992: 38).

To begin to move beyond this quandary, Wynne contends, it is necessary to dispense with the assumption that a lack of public identification with science can be equated with the public lack of understanding of science. This implied relationship of equivalence, as noted above, formulates the main issue at stake as one of inducing scientists to find new ways of making science more popular or entertaining for members of the 'lay' public. Even more seriously, as Wynne (1992: 38) points out, questions 'such as those about whose interests are served by different kinds of science and scientific representation, and about the basis of trust and social accountability of different institutional forms of control and ownership of science, are effectively deleted'. Consequently, by centring for investigation precisely these kinds of questions, critical analyses can interrogate what are otherwise unacknowledged factors shaping the public uptake or 'understanding' of science. In this way, by opening up for analysis the politicized ways in which people experience science socially, researchers will be better placed to engage with what Wynne and others perceive to be the gradually deepening crisis of science's public credibility and authority (see also Irwin 1995; Irwin and Wynne 1996; Lash et al. 1996; Adam et al. 2000; Nieman 2000; Irwin 2001).

Science on display

In order to round out this chapter's discussion, our focus shifts to the ways in which people experience science socially at a particular range of institutional sites, namely science museums and centres. 'The construction of a museum', as Morton (1988: 130) maintains, 'is a signal the government and others want to influence public attitudes to science and technology, to increase the standing of these subjects (and that of their practitioners in particular).' Questions regarding how best to put science on display, so to speak, raise complex issues far beyond the more immediate ones of institutional investment, management and marketing. Despite being typically presented as situated 'above' politics, science exhibitions have frequently been the subject of intense ideological conflicts over what constitutes 'science' and how to communicate it to 'the public'. In the words of one museum director: 'In those choices of what to present or what not to present, we really do, unwittingly sometimes and consciously other times, frame a moral point of view for our visitors about science and about science topics' (Sullivan 1992: 128).

These types of ideological conflicts over how best to present 'the facts of the natural world' emerged, not surprisingly, with the arrival of the earliest science exhibitions. Histories of the origins of these exhibitions often differ over which one can legitimately claim to be the world's first, mainly because of disputes over precisely which criteria should be invoked when making such a judgement. That said, most historical accounts concerning the advent of the museum in early modern Europe maintain that it was a seventeenth-century innovation, in effect the product of the first cabinets of 'natural curiosities' and 'rarities' gathered by merchants and explorers in the course of their great voyages of discovery. The first public museum in Britain, the Ashmolean Museum in Oxford, drew on John Tradescant's *Musaeum Tradescantianum* (based on the collection of his father, a naturalist and diplomat). Opening in 1683, it 'celebrated the new scientific outlook of the Renaissance', being intended for 'the knowledge of Nature' as acquired through 'the inspection of Particulars' (Hackman 1992: 65; Quin 1993). The methods of collecting antiquities, Hackman (1992) suggests, were being applied to the natural world.

The public museum began to acquire its modern form towards the end of the eighteenth century, a process of formation which owed much to the development of other kinds of institutions – such as the international exhibition and the department store (Bennett 1995: 19). The British Empire's 'Great Exhibition' of 1851, covering 21 acres in South Kensington, London, attracted an extensive array of objects, many of which were eventually organized into a museum in 1857. 'By 1900,' according to Pyenson and Sheets-Pyenson (1999: 134), 'Germany could claim 150 natural history museums, Britain 250, and France 300, while the United States counted a respectable 250.' London's Science Museum was established in 1928, its aim being to 'record achievements in science, and by displaying these in the form of technological objects, to inspire further efforts: it should "serve to increase the means of Industrial Education and extend the influence of Science and Art upon Productive Industry"' (Gregory and Miller 1998: 198). In the US, 'world's fairs' played a similarly decisive role in shaping public images of science as they sought to 'record the world's advancement' in terms of technological progress and public enlightenment.

For science museums today, the cultural politics associated with putting science 'on display' continue to pose acutely difficult problems of representation. It is important to acknowledge, as Kavanagh (1992: 82) notes, 'that no museum is innocent, that museums make visible our myths, that they hold the stories we tell ourselves about ourselves, and they exist in a world saturated by ideology and therefore cannot be neutral'. Accordingly, she maintains that exploring science museums 'for their hidden agendas, their unwitting testimony, their coded, partial (often male) views of the world, is improving our appreciation of how the museum is constructed' (1992: 82). Such critical lines of inquiry have informed a range of interesting studies into how museums create particular kinds of science for the public. Macdonald's (1996, 1998) research, for example, based on her fieldwork at the Science Museum in London, leads her to observe:

> science communication involves selection and definition, not just of which 'facts' are presented to the public, but of what is to count as science and of what kind of entity or enterprise science is to be. That is, science communicators act as authors of science for the public. They may also, however, by dint of their own institutional status, give implicit stamps of approval or disapproval to particular visions or versions of science. That is, they may act as authors with special authority on science – as authorisers of science.
>
> (Macdonald 1996: 152)

This recognition of how museums perform the role of 'authoring' or 'authorizing' science leads, in turn, to a range of issues concerning what their visitors want or need to know. Seemingly 'common sensical' assumptions about scientific knowledge, as well as how it may be most effectively expressed via exhibitionary forms and techniques, can be shown to have a deeply political significance upon closer inspection. 'Not only do science communicators define science for the public,' Macdonald (1996: 153) writes, 'they also in effect build a vision of "the public", and the kind of "understandings" that the public can be expected or hoped to make, into their communications.' Many of the debates associated with science museums concern their relative capacity to improve the public 'uptake of science' by making it appear more engaging, worthwhile and accessible to 'lay' people. Such efforts are usually assumed to culminate in a bolstering of popular support for science and scientists, at times in the face of evidence to the contrary.

In any case, members of the public visiting science museums are likely to find themselves being addressed in at least two distinctive ways. First, in political terms – that is, as citizens who need to know about science so as to participate effectively in a democracy – and, second, in economic terms – that is, as consumers being encouraged to purchase the museum experience, likely as a leisure activity. Macdonald's (1998) account of how these and related definitions of the public inform a science museum's vision of its visitors, as well as of the science on display, leads her to pinpoint a series of important questions:

- Who decides what should be displayed?
- How are notions of 'science' and 'objectivity' mobilized to justify particular representations?
- Who gets to speak in the name of 'science', 'the public' or 'the nation'?
- What are the processes, interest groups and negotiations involved in constructing an exhibition?
- What is ironed out or silenced?
- And how does the content and style of an exhibition inform public understandings?
- Who is empowered or disempowered by certain modes of display?

(Macdonald 1998: 1, 4)

In seeking to discern how a science exhibition imagines its audience, Macdonald proceeds to raise questions about who is included by also who is excluded from its preferred definition. Similarly, while recognizing that visitors will redefine its significance largely in their own terms, she also questions the extent to which exhibitionary strategies enable certain kinds of readings at the expense of alternative ones.

Interestingly, in the course of questioning visitors to a particular Science Museum exhibition concerned with food, Macdonald is able to gain insights into what counts as 'scientific' in their eyes. Her interviews with them suggest that several of them actively discerned distinct 'levels of scientific-ness'; that is, some differentiated between 'real science' or 'pure science' and 'popular science'. This 'partitioning' of science, she maintains, appeared to indicate that the exhibition succeeded in challenging the preconceptions of visitors, yet rather worryingly 'science' proper *per se* 'was sometimes still moved elsewhere, still out of the public's grasp' (Macdonald 1996: 165). Consequently, this research indicates to her that science 'can be difficult and distant, it can be gendered and racist, it can be hedged about with all kinds of barriers and vested interests,' and that understanding this is 'public understanding of science' too (1996: 168). At the same time, though, 'the case study suggests that enlisting support to promote such understandings as these would not be easy within the contexts within which a national museum must operate' (Macdonald 1996: 168; see also Silverstone 1992).

Science centres are usually differentiated from science museums on the basis of their provision of 'hands on' experiences. Historians typically point to the formative influence of Francis Bacon's *New Atlantis*, published in London in 1627, with its imaginary House of Salomon wherein:

> we make demonstrations of all lights and radiations, and of all colours [. . .] We procure means for seeing objects afar off; as in heaven and remoter places . . . We also have helps for sight, far above spectacles and glasses in use . . . We also have sound houses, where we practice and demonstrate sounds and their generation. We have harmonies which you have not, of quarter-sounds and lesser slides of sounds . . . We have the means to convey sounds in trunks and pipes, in strange lines and distances . . . We imitate also flights of birds . . . we have ships and boats for going under water, and brooking of seas; also swimming girdles and supporters. We have divers curious clocks, and other like motions of return, and some perpetual motions . . . We have also houses of deceits of senses; where we represent all manner of feats of juggling, false apparitions, impostures, and illusions; and their fallacies. And surely you will easily believe that we have so many things truly natural which induce admiration, could in a world of particulars deceive the senses, if we could disguise those things and labour to make them seem more miraculous.
>
> (cited in Gregory and Miller 1998: 200–1)

This conception of a new type of institution devoted to the pursuit of knowledge about the natural world was quite remarkable. This was especially so, as the above account makes clear, with respect to how different technologies – microscopes, telescopes, as well as possible hints of aeroplanes, submarines and telephones – might be presented to its visitors for purposes of experimentation. Bacon's imaginative portrayal of a House of Salomon continues to be cited to this day as a manifesto of sorts for interactive science centres (Gregory and Miller 1998).

Science centres typically assume as a principal aspect of their remit the task of stimulating public curiosity about science. This aim is to be achieved, for the most part, by inviting the visitor to engage with science via the various interactive forms of exhibits and demonstrations. Indeed, in some ways the success of the science centre may be read as a response to a criticism frequently made of science museum displays, namely that they 'are about technology, not science, mainly because it is so much easier to present technological artifacts than abstract scientific ideas' (Farmelo 1998: 353). In the case of Science City in Calcutta, India, for example, 'interactive exhibits and hands-on activities provide a viable alternative to other science popularisation programmes'. In the words of its director, the centre is 'an environment without run-of-the-mill displays and gives a never-enjoyed-before opportunity to step into a new world and find out the scientific principles behind so many things'. At the Ontario Science Centre in Canada, an effort is made to 'convey the excitement felt by scientists as they break through to new discovery'. Techniquest in Cardiff, Wales seeks to ensure that 'visitors are positively encouraged to use all their senses to explore the fascinating world of science for themselves'. Meanwhile, at Explore-at-Bristol in England, 'the exhibitions and events provide stimulating starting points, raising questions, promoting creativity and encouraging citizenship.' Science centres such as these ones unashamedly dispense with the norms of the museum by striving to develop 'participatory' strategies that inspire fresh ways of thinking about science and technology. Moreover, they open up new possibilities for social inclusion and lifelong learning (see also POST 2000b).

While acknowledging the success of some science centres in meeting these and related objectives, Bradburne (1998) maintains that most centres are insufficiently self-reflexive about their own practices. Typically it is the case, he argues, that recently constructed science centres continue to be based on fairly traditional exhibitionary approaches, despite their limited success in communicating science. Such approaches, in his experience, share three weaknesses: 'They focus almost exclusively on principles and phenomena rather than processes, they misrepresent the nature of scientific activity, and they show science out of context – science defined "top down" by scientists, rather than as experienced by visitors' (Bradburne 1998: 238). Even where efforts are made to situate science and technology within a social context,

Bradburne claims, attention usually remains focused on science and technology at the expense of the larger society. The root of the problem, he suggests, is that the 'dominant model in which science centers "vulgarize" knowledge and make it available to the masses, or sugar-coat science with gratuitous hands-on interaction to arouse visitor curiosity, is rarely if ever questioned' (1998: 239). Accordingly, science centres need to undergo a profound transformation, in his view, so as to expand the opportunities available to reach audiences much more interested in society (here he cites issues such as environmental protection, genetic manipulation, euthanasia, urban development and crime) than in science and technology as such. Science centres, it follows, cannot afford to be defined as standing apart from 'culture as a whole'. Bradburne insists that they must acquire the flexibility to speak to the changing needs of a wide variety of users, and in a manner which helps to support 'self-initiated, self-directed, and self-sustained learning in an informal setting' (1998: 245).

Responses to Bradburne's intervention, such as that of Persson (2000), are typically much more upbeat about the future prospects of the science centre in its current (albeit still evolving) form. For the most part, supporters insist, such institutions are thriving, with overall public attendance steadily growing worldwide. Many science expositions now rank as top tourist attractions in major cities, their popular appeal – it is frequently claimed – allowing for substantive inroads to be made into the promotion of scientific literacy, especially among young people. Encouraged by this sense of progress, many supporters are similarly excited about the potential to develop even more effective types of display and interactivity through the innovative use of new media technologies, not least via the Internet (see also Mellor 2001a). In thinking of such sites as 'arenas for public debate', Henriksen and Frøyland (2000: 394), for example, are interested in how interactive exhibits can 'facilitate meaningful dialogues and conversations among visitors' so as to help realize the science museum's potential contribution 'to the civic and practical aspects of scientific literacy'. In bringing this chapter to a close, however, I wish to turn to Shamos's (1995) critical assessment of the very notion of 'scientific literacy' itself.

In *The Myth of Scientific Literacy*, a book written on the basis of his experience as both a physicist and science educator, Shamos launches a provocative challenge to the 'ever-elusive goal' of universal scientific literacy. The 'notion of developing a significant scientific literacy in the general public', he writes, 'is little more than a romantic idea, a dream that has little bearing on reality' (Shamos 1995: 215). In the case of science museums and centres, he readily acknowledges that the 'actual hands-on experience' of science and technology proffered by some exhibitions has considerable appeal among children and adults. Indeed, for young people, the possibility of a career in science may seem more attractive as a result. At the same time, though, he points out that such sites are still relatively scarce, even in the US despite longstanding government initiatives aimed at enhancing scientific literacy, and that the institutions themselves are costly to establish and maintain. Moreover, and herein lies the rub, he questions the extent to which one can actually expect a reasonable share of the adult population to be sufficiently interested in science to bother attending such sites in the first place.

Science, Shamos maintains, is invariably presented to the public 'ex post facto, so to speak, as a fait accompli, with facts, theories and laws already tied into a neat package to be opened, examined, and admired (or scorned) by the prospective learner' (Shamos 1995: 218). For this and related reasons, he argues, few members of the general public take an active interest in scientific issues at all, while only a small fraction of them might be considered 'scientifically literate' by even the most general definition. Consequently, Shamos contends that the notion of 'scientific literacy' should be discarded as a meaningless goal and replaced, in turn, with the term 'science awareness'. Such an approach, it follows, would recognize science as a cultural imperative, that is, as a method of inquiry as opposed to a 'tried and true' routine for acquiring knowledge of the universe. In this way, fresh appraisals may be undertaken into how different publics are made aware of 'science in the making: what goes into preparing these packages, what science does, and how and why it goes about its practice' (1995: 218).

This move to recast the objective of scientific literacy in more democratic terms clearly invites a thorough reconsideration of familiar assumptions about how factual representations of science are being inflected more widely across the public sphere. Beginning in the next chapter, our focus will turn to science journalism so as to ask: who defines what counts as 'newsworthy' science, under what conditions, and why? Questions will be similarly posed about the influence news media images of science have in shaping public perceptions of the risks associated with today's scientific and technological controversies for tomorrow. 'Science's uncertain future needs to be presented as just that,' observes Hornig Priest (1999: 111), 'the unknown and the unknowable that we must nevertheless consider, lest it arrive before we are ready'.

Further reading

Best, S. and Kellner, D. (2001) *The Postmodern Adventure*. London: Routledge.
Broks, P. (1996) *Media Science before the Great War*. London: Macmillan.
Dawkins, R. (1998) *Unweaving the Rainbow*. Harmondsworth: Penguin.

Dunbar, R. (1995) *The Trouble with Science*. London: Faber and Faber.
Gould, S.J. (1996) *Dinosaur in a Haystack*. London: Jonathan Cape.
Haynes, R.D. (1994) *From Faust to Strangelove: Representations of the Scientist in Western Literature*. Baltimore, MD: Johns Hopkins University Press.
Irwin, A. and Wynne, B. (eds) (1996) *Misunderstanding Science? The Public Reconstruction of Science and Technology*. Cambridge: Cambridge University Press.
Labinger, J.A. and Collins, H. (eds) (2001) *The One Culture?* Chicago: University of Chicago Press.
Nowotny, H., Scott, P. and Gibbons, M. (2001) *Re-Thinking Science*. Cambridge: Polity.
Ross, A. (ed.) (1996) *Science Wars*. Durham, NC: Duke University Press.
Sagan, C. (1997) *The Demon-Haunted World: Science as a Candle in the Dark*. London: Headline.
Shamos, M.H. (1995) *The Myth of Scientific Literacy*. New Brunswick, NJ: Rutgers University Press.
Ziman, J. (2000) *Real Science*. Cambridge: Cambridge University Press.

SCIENCE JOURNALISM

> I've heard scientists mourn the good old days, when the public believed what scientists said. I believe that blind faith is dangerous. Today's skepticism about science comes, in part, from a sense of betrayal: 'We believed you when you said that science would perfect the world. And look at it now.' By reporting on science in all its dimensions, we make it real. In doing so, in letting people inside the process, it is true that we might decanonize the scientific society. But we also bring science back into the real world – where it belongs.
>
> (Deborah Blum, science journalist)

Science, it is often said, gets a bad press. Explanations for this apparent problem, in the opinion of some journalists at least, tend to revolve around the charge that most types of science fail the test of newsworthiness. Routine science, they tend to believe, is really rather boring. It lacks the stuff of drama necessary to spark lively newspaper headlines. At the same time, some scientists maintain that on those occasions when a certain scientific development is given due prominence, it all too frequently happens for the wrong reasons. Not surprisingly, they are quick to condemn instances of sensationalist reporting – where news values have given way to entertainment values – for misrepresenting the nature of scientific inquiry, and rightly so.

'Media wisdom has it', writes Dunbar (1995: 147–8), 'that news must have impact and, especially, human interest to sell papers. But when science tries to compete with the social antics of the great and the not so good, it has only limited chances of success.' Accordingly, to critically engage with the most urgent issues at stake in science journalism is to acknowledge from the outset the uneasy tensions which exist between discourses of science and those of journalism. The phrase 'science journalism' can be misleading in its apparent claim to characterize a specific kind of reporting, implying as it does that the boundaries between it and other types of journalism are both well marked and dutifully respected. As even a cursory glance at a newspaper or television newscast reveals, however, this is anything but the case in practice. Scientific endeavours are typically highly complex affairs, and any attempt by journalists to impose narrative order on them in the name of a 'good story' is to necessarily risk engendering compromises. Precisely how the threads making up the fabric of science reporting are interwoven in a given news account is as much a matter of subjective interpretation of 'what's at stake for society' as it is of any self-proclaimed 'objective' methods of reporting.

This reference to the fabric of science reporting is a useful one for our purposes here. Over the course of this chapter, I shall aim to give several of its dangling threads a firm tug. It is my hope that in doing so some of the unspoken assumptions underlying ongoing debates about science journalism will come unravelled so that we may be better placed to understand – and challenge – them in the chapters to follow.

Science and society

'Society's relationship with science is in a critical phase,' declared a report by the British House of Lords Select Committee on Science and Technology in March 2000. The report, titled simply *Science and Society*, drew upon evidence collected over a yearlong inquiry into the widespread perception that there is a serious crisis of public confidence in the biological and physical sciences and their respective technological applications. In the course of presenting its findings, the report identifies a range of issues which resonate, to varying degrees, in public debates across a range of different national contexts. As such, it will be used here as a springboard of sorts for the discussion to follow.

Underpinning this apparent crisis in public confidence is a paradox. At one level, the report states, 'there has never been a time when the issues involving science were more exciting, the public more interested, or the opportunities more apparent' (Select Committee on Science and Technology 2000). These claims are supported with reference to the results of recent opinion survey studies, as well as with regard to the growing salience of popular media output addressing scientific topics, among other factors. Nevertheless, on another level, 'public confidence in scientific advice to Government has been rocked by a series of events, culminating in the BSE [or "mad cow disease"] fiasco; and many people are deeply uneasy about the huge opportunities presented by areas of science including **biotechnology** and information technology, which seem to be advancing far ahead of their awareness and assent' (Select Committee on Science and Technology 2000). Precisely how this enhanced level of public interest corresponds, in turn, with what the report describes as an 'increasing scepticism about the pronouncements of scientists on science-related policy issues of all types' is the subject of much debate.

In the view of the select committee's members, which included several distinguished scientists under the chairpersonship of Lord Jenkin of Roding, 'public unease, mistrust and occasional outright hostility are breeding a climate of deep anxiety among scientists themselves' (2000). Interestingly, the report signals from the outset its commitment to exploring anew the varied sources of these tensions. Identified as being one of the more influential sources, as might be anticipated, is science journalism. Public attitudes to science, the report points out, are shaped by an array of institutions situated across the breadth of society, not least by the teaching of science in schools. Once people leave the education system, however, the news media become their principal sources of information about science. Therein lies the problem in the eyes of scientists. According to the select committee report, many scientists tend to be convinced that journalists 'have it in for them', that is, that the 'cherished freedom of the British press works against them'.

In order to unravel the complexities of science journalism, the committee distinguishes between three different types. First, there is the specialist scientific press, where news reports are written by scientists for other scientists. Second, there is the work of science journalists, namely specialist correspondents employed by mainstream news organizations. They will usually conduct their own journalistic research into a science story so as to ensure due accuracy in their handling of the facts. Third, there is the work of non-scientific correspondents. These journalists, namely by dint of circumstance, typically find themselves writing about a scientific development as a general news story. In so doing, the report maintains, they may subject the story to 'a very different set of values and criteria' than might be otherwise expected from a specialist science reporter.

In comparing the activities of the latter two types of journalists, the report highlights several factors which together help to explain why the information about science being presented by the news media is not as effective as it might be. These factors begin with the simple observation that science reporters are first and foremost concerned with their role as a journalist, as opposed to that of an educator: 'Their primary aim, as with any journalist, is to get stories into the paper or programme, in fierce competition with other journalists.' This determination to see the scientific world in terms of news stories leads, in turn, to an undue emphasis being placed on those kinds of events that satisfy the news organization's editorial policy.

Science and news, the report suggests, tend to be 'a poor fit', namely because 'newsrooms deal in simplified stories put together in haste, preferably with two opposing sides or views'. One witness before the committee is cited as recalling the following instructions of a BBC Radio 1 news producer prior to a live interview: '20 seconds, professor, and no long words.' Moreover, when priority is given to clashing viewpoints, especially when they are sensationalist ones, familiar notions of balance and fairness can be decisively undermined. The committee heard 'vehement criticism' of the practice, for example, where the news media 'give equal weight to the scientific consensus and to a minority view, whether in the interest of balance as they see it, or simply because confrontation makes good copy'. Particularly revealing is the report's contention that this crisis is at its most severe where scientific definitions of 'risk' are concerned.

'When science and society cross swords,' the report intones, 'it is often over the question of risk.' As it proceeds to point out, there are two dimensions to risk which are particularly significant in this context: 'the chance of something happening, and the seriousness of the consequences if it does.' How scientists choose to communicate their calculations of risk is not only a question of rigour and accuracy, but also one of politics. Acute difficulties may arise, for example, as soon as scientists undertake to formally quantify a risk, especially when they necessarily have to qualify their claims by acknowledging a degree of uncertainty ('It appears to be safe, on the basis of the following assumptions which require further research'). Confronted with such hedged assurances, journalists will more likely than not insist upon absolute certitude ('Is it 100 per cent safe?'). Their inability to extract such an affirmation from scientists can sometimes lead them, in turn, to call into question the integrity of the actual risk calculation itself on these grounds alone. 'By this means,' the report suggests, 'the stage is set for confusion, cynicism and even panic.'

Not surprisingly, various individuals and groups seeking to help secure a place for these issues on the national agenda have warmly welcomed this report. Many publicly applauded its commitment to recasting several of the all too familiar premises underpinning debates about the science–media nexus. Finding particular support among those who regard the status quo as untenable was the committee's conviction that 'the culture of UK science needs a sea-change, in favour of open and positive communication with the media.' To date the government's response to the report has been positive, if largely limited to a formal reaffirmation of its importance as a contribution to ongoing debate. In particular, the report has been credited with influencing the development of policies set out in the Science and Innovation White Paper, published in July 2001. Disappointingly, if all too tellingly, the report's publication was virtually ignored by the very news organizations it set out to challenge. In the case of the broadsheet newspapers, for example, coverage typically consisted of a single item acknowledging the report's release, together with a brief sketch of its contents. It was then promptly consigned to the journalistic dustbin of history.

My decision to dwell on the report's engagement with science journalism here is based on my sense that it succeeds – both by accident and by design – in effectively highlighting several pressing issues in need of further attention. Reading the report 'against the grain', so to speak, it is possible to discern an array of tacit, seemingly 'common sensical' assumptions which take on a distinct ideological significance upon closer scrutiny. In the remaining portion of this section, I would like to pinpoint four sets of questions revolving around a specific aspect of the report's claims. Each cluster is intended to bring to the fore conceptual issues for further discussion in the remaining portions of the chapter.

Discourses of 'the public'

The public is becoming increasingly distrustful of science-based expertise, the report warns. What happens to this thesis if 'the public', understood as a singular, cohesive totality, is made to give way to a pluralized conception of publics? To recognize the tensions engendered by various discourses of 'the public' is, at the same time, to acknowledge that a multiplicity of publics exist where the report recurrently posits only one. If, as Williams ([1958] 1989: 11) once famously observed, 'there are in fact no masses, only ways of seeing people as masses', then configurations of 'the public' similarly need to be scrutinized precisely as they support, overlap and contradict one another. In attempting to discern the hierarchical relations of social power shaping the mobilization of these competing configurations, questions such as the following are key. Specifically, who claims to speak on behalf of 'the public', who is identified as representing 'public opinion', and who falls outside of such definitions, thereby possibly threatening the 'public interest'?

Trust in science

A guiding principle of the report is that efforts to promote greater 'trust in science' are consistent with the basic requirements of modern society. Do journalists, as citizens, have a responsibility to report on science in such a way as to rebuild public trust in it (in parallel with scientists who, the report suggests, should see themselves as 'civic scientists')? The report answers this question in the affirmative, thereby indicating its underlying dependence on

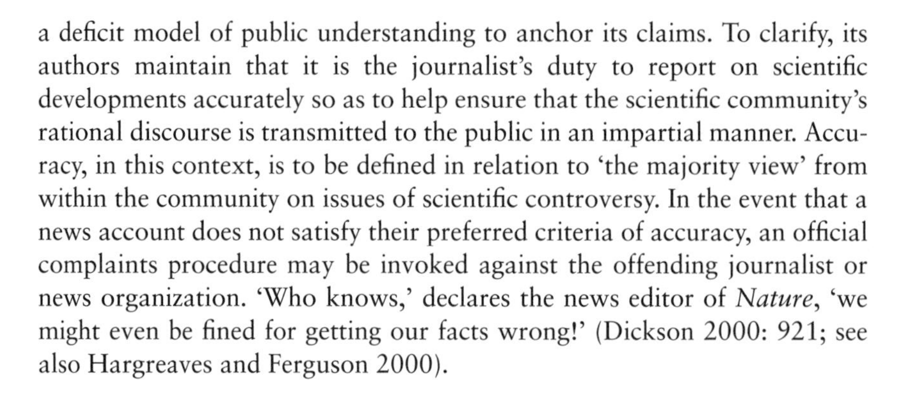

a deficit model of public understanding to anchor its claims. To clarify, its authors maintain that it is the journalist's duty to report on scientific developments accurately so as to help ensure that the scientific community's rational discourse is transmitted to the public in an impartial manner. Accuracy, in this context, is to be defined in relation to 'the majority view' from within the community on issues of scientific controversy. In the event that a news account does not satisfy their preferred criteria of accuracy, an official complaints procedure may be invoked against the offending journalist or news organization. 'Who knows,' declares the news editor of *Nature*, 'we might even be fined for getting our facts wrong!' (Dickson 2000: 921; see also Hargreaves and Ferguson 2000).

Mediating science

In discussing the imperatives of science journalism, the report takes issue with how the scientific world is reflected in news accounts. This language of reflection is employed to pinpoint evidence of the extent to which journalists allow certain distortions to creep into the reporting process. Such an approach, I would argue, fails to account for the ideological dynamics embedded in the reporter's mediation of scientific controversies. That is to say, while journalists typically present a news account as a *translation* of reality, it may be better understood to be providing an ideological *construction* of contending truth-claims about reality. To deconstruct a news accounts in ideological terms, then, it is necessary to ask: why are certain truth-claims being framed by journalists as reasonable, credible and thus newsworthy while others, at the same time, are being ignored, trivialized or marginalized? Responses to this question will help bring to the fore the truth-politics of science reporting in all of their complexities.

The limits of criticism

Consistent with current debates around the public understanding of science, much is made of a 'dialogue' approach in the report. The assumption is that an enhanced culture of openness about all aspects of publicly supported scientific research will help to restore public trust in science. It follows, however, that the relative degree of this openness must nevertheless respect certain normative limits. Criticisms of science, the report intones, 'whether well-founded or misguided' on the part of the public, 'may inhibit technological progress'. Responsible criticism, by this logic, is criticism that avoids calling into question the legitimacy of technological progress by reporting views 'held by only a quixotic minority of individuals'. Implicit here is the precept that technological progress will ultimately translate, by definition, into social progress (Miller 2001). It follows that voices seeking to call into question the apparent inevitability of this equation will risk being defined as anti-science – and hence irrational at best, or dangerous at worst (see also Dickson 2000).

Bearing these issues in mind, we turn in the next section to explore the factors shaping how news organizations determine whether a scientific issue is sufficiently 'newsworthy' to warrant public attention.

Making science newsworthy

'Since the earliest days of science writing,' Friedman (1986: 17–18) observes, 'the profession has been beholden to the two worlds of science and journalism, functioning under the rules and constraints of both.' Moreover, she adds, 'the influence of these two worlds has not been equal', with those involved from the journalistic side playing a more decisive role in shaping the development of science writing to meet their respective needs. This contest continues today, of course, as the status of science writing as a distinct strand of journalism is being increasingly recognized across a range of media genres.

'Over the last 30 years or so,' as science reporters Blum and Knudson (1997: ix–x) point out in a US context, 'science writing has been transforming itself into something beyond a strange little subculture of journalism', namely as a profession in its own right (see also Friedman et al. 1986; Dornan 1988). Driving science writing as a profession, in their view, is first and foremost science itself:

> the post-World War II boom in research, the space race of the 1960s, the technologies of today that are opening the subatomic and molecular worlds at a still-dizzying pace, giving rise to a revolution in personal communications and in our knowledge of genetics and biology. Such discoveries have altered the world we all live in, and it has increasingly fallen to the media to explain the new technologies and report on their impact, good and bad.
>
> (Blum and Knudson 1997: x)

Where science journalists once acted primarily as 'scouts' on a reconnaissance mission, 'trying to bridge a big divide by bringing back messages from one side to the other', today they are unlikely to be content to limit their work to this role alone. Although they still need to perform the task of translating the abstract complexities of scientific inquiry for their audience, Blum and Knudson (1997: x) maintain that the science reporter's role must also stretch to encompass a larger sense of public responsibility. In their words:

> You can paint an awesome and adventuresome picture of space exploration with all its glittering planetary rings, but you can also acknowledge its risks and probe its failures. You can point out the medical and agricultural benefits of the new biotechnology or the mapping of the human and other **genomes**; but you can also question what harm may come of the new knowledge and capabilities, discuss what safeguards will be put in place, and talk about how much big science costs and who pays for it.
>
> (Blum and Knudson 1997: x)

It is this shift from 'translating' science for members of the public so as to, in turn, render problematic its underlying precepts that is crucial in this regard. Science writing, if for many of its practitioners the most deeply satisfying form of journalism to learn, is one of the most difficult to get right.

Science typically appears in the press as 'an arcane and incomprehensible subject', Nelkin (1995) observes, around which there is a certain 'mystique' that implies it is to be properly regarded as a 'superior culture' with a 'distanced and lofty image'. Far from enhancing public understanding, she argues, 'such media images create a distance between scientists and the public that, paradoxically, obscures the importance of science and its critical effect on our daily lives' (Nelkin 1995: 15; see also Dutt and Garg 2000; Shachar 2000). Indeed, of the various beats newspaper reporters regularly cover, the science beat is one of the most challenging. 'In most other speciality beats,' as Rensberger (1997: 8) of the *Washington Post* points out, 'reporters become familiar with a modest body of knowledge (how a city council functions, for example, or the rules of baseball) and turn to the same few, first-name-basis sources every day.' The science reporter, he adds, seldom enjoys this luxury, having instead to come quickly up-to-speed on a host of emerging events or issues as they surface from one day to the next. Breaking science news is very difficult to anticipate: 'Today the story may be a claimed advance in treating cancer, tomorrow it may be explaining atmospheric chemistry and, for the weekend, the latest experiment in fusion power research' (Rensberger 1997: 8). As quickly as the issue changes, of course, so will the reporter's perception of which potential sources of information are likely to be most appropriate for the news item being prepared.

One of the telling realities of writing about science for a daily newspaper, according to Rensberger, can be found in the mailroom. There, he argues, one will typically find that the 'mailboxes of the science and medical reporters will be among the most stuffed' (Rensberger 1997: 7). Demands on the science reporter's time and attention are as constant as they are vociferous. In Rensberger's experience, such demands are likely to arise from individuals and groups in all walks of life, but especially from people associated with 'universities, corporations, think tanks, government agencies, advocacy groups, independent research institutions, museums, public relations agencies, hospitals, and scientific journals' (1997: 7). The sheer volume of these efforts to capture the newspaper science reporter's interest necessarily means that she or he has to assume a 'gatekeeper' function in the newsroom. That is to say, the science reporter will have to decide on a routine basis 'what developments in the real world get into the news, and hence reach the public'. This gatekeeping function is obviously fraught with difficulties, not least where the reporter's proclaimed objectivity is concerned. 'After all,' Rensberger (1997: 11) argues, 'of the dozens of stories we could do on any given day, we reject most or all possibilities. In this we exercise our opinion as to what is a good story' (see also Saari et al. 1998; Kiernan 2000; Malone et al. 2000).

And what makes a good science story, in Rensberger's opinion? In essence, he maintains that the following five criteria – in combination, and to varying degrees from one story to another – are especially valued by newspaper science journalists:

- *Fascination value*: Rensberger (1997: 11) writes: 'This is the special commodity that science stories, more than any other kind, have to offer. People love to be fascinated, to learn something and think, "That's amazing, I didn't know that".' By this criterion, then, dinosaurs 'may be the quintessential fascinating topic for science writers'.
- *Size of the natural audience*: here Rensberger is referring to the number of newspaper readers who are already aware that they are interested in following a news story about a given topic. 'If the story is about a common disease that everyone had had or fears getting,' he declares, 'the natural audience will be larger than for a rare disease' (1997: 11).
- *Importance*: the subjective quality of any attempt to assess importance is readily acknowledged by Rensberger, although he suggests that to 'judge a story idea on this point, you would try to decide whether the event, or finding, or wider knowledge of the event or finding is going to make much of a difference in the real world, especially in that of the average newspaper reader' (1997: 11–12). Following this logic, 'AIDS is important, bunions are not.'

- *Reliability of the results*: to pinpoint this criterion, Rensberger poses the question: 'Is it good science?' The single most useful guideline for determining reliability, he argues, is science's own peer review system. This system, he writes, 'is a time-tested way to minimize the odds that a misunderstanding is promulgated to the world at large. Science writers who ignore the system risk misleading their readers and embarrassing themselves' (1997: 12; see also Young 1997).
- *Timeliness*: 'The newer the news,' Rensberger (1997: 13) states, 'the newsier it is.'

Staying with this last point about timeliness, it is interesting to note that the temporal specificity of the previous four factors is also significant in this regard. That is, their timeliness becomes apparent when considering some of the additional examples Rensberger gives to further illustrate his criteria, such as his claim that newspaper readers will attach a much higher fascination value to black holes over meteors. Today it is much more likely to be the other way around. Fears about the potential risk of a giant **meteorite** obliterating life on Earth are receiving extensive coverage in the news media, as well as treatment in Hollywood films such as *Deep Impact* (1998), *Armageddon* (1998) and the like (see also Mellor 2001c). Similarly, his claim that Creutzfeldt–Jakob disease registers at the low end of the scale with regard to the size of its 'natural audience' is, sadly, no longer the case due to public anxieties about the human form of BSE or 'mad cow disease'.

The processes of selection indicative of one news organization will be at variance with those of others, of course, but shared assumptions about these and related criteria of 'newsworthiness' recurrently underpin these daily negotiations. News coverage of science tends to favour certain areas of scientific inquiry over and above other areas, a pattern which is usually observable from one newspaper or news broadcast to the next. The process by which certain scientific developments are rendered 'newsworthy' while others, in contrast, are deemed unworthy of attention, is the outcome of a complex array of institutional imperatives. Journalists, together with the other individuals involved in the work of processing news in a particular news organization (editors play a key role here), bring to the task of making sense of the social world a series of 'news values'. These news values are operationalized by each newsworker, as Hall (1981) suggests, in relation to their 'stock of knowledge' about what constitutes 'news'. If all 'true journalists', he argues, are supposed to know instinctively what news values are, few are capable of defining them:

> Journalists speak of 'the news' as if events select themselves. Further, they speak as if which is the 'most significant' news story, and which

'news angles' are most salient, are divinely inspired. Yet of the millions of events which occur every day in the world, only a tiny proportion ever become visible as 'potential news stories': and of this proportion, only a small fraction are actually produced as the day's news in the news media.

(Hall 1981: 234)

Hence the need to problematize, in conceptual terms, the operational practices in and through which news values help the newsworker to justify the selection of certain types of events as 'newsworthy' at the expense of alternative ones. To ascertain how this process is achieved, researchers have attempted to explicate the means by which certain news values are embedded in the very procedures used by reporters to impose some kind of order or coherence on to the social world (see also Allan 1999).

Hansen (1994: 114–15), drawing on his study of science reporting in the British press, suggests that the 'most pronounced criterion of newsworthiness is whether science can be made recognizable to the reader in terms of human interest or in terms of something readers can relate to'. Particularly prized, as a result, are those events which illuminate the relevance of science to daily life, enabling the journalist to adopt a 'human angle' when constructing the news story. Further factors informing this process of negotiation include the efforts made by news sources or **stakeholders** themselves to influence journalistic judgements, as well as the relative complexity of the event itself. 'The more complex or inaccessible a piece of science news is,' Hansen (1994: 115) writes, 'the more "translatory work" it will require on the part of the journalist to make it intelligible and interesting to the readers.' Time is of the essence for journalists working to conform to a daily production schedule, especially where deadlines are concerned. That said, while a significant scientific 'breakthrough' may be judged to constitute 'hard' news, and thereby warrant immediate coverage, it is much more likely that the science involved will be, 'in news terms, a slow process of small incremental developments' (1994: 115). The 'event-frequency' of science, Hansen contends, seldom corresponds with the 'news-frequency' of the reporting time-cycle:

'Science' *per se* is not seen as hard news, but remains part of what some journalists call the 'soft underbelly' of news coverage, and science stories tend to get squeezed out if set in direct competition with other, more mainstream types of news, particularly political news. With the exception of the 'weird-and-wacky' or the 'implications-for-the individual' types of science, science predominantly becomes 'news' (i.e. moves from the domain of the specialist science sections within a

> newspaper to the main news sections) if it is linked to major developments in the political or economic sphere.
>
> (Hansen 1994: 116)

Science, it follows, usually will be the main focus of a news story only when it is directly linked to a wider issue or problem, especially one that has unfolded over the previous 24 hours. Far more typically, science provides a supplementary thematic, a so-called 'soft' news or 'human interest' item that is much less dependent on notions of 'timeliness' than is the case with 'hard' news items.

News values are culturally specific, and never fixed once and for all – instead they are always evolving over time, and are inflected differently from one news organization to the next. Still, the ostensibly 'common sensical' criteria informing definitions of newsworthiness in science journalism have proven to be surprisingly consistent over the years. 'For all of modern science's sophisticated concepts and technology,' Young (1997: 114) observes, 'journalism's traditional five *Ws* and an *H* – who, what, when, where, why and how – still form the core of science reporting.' Toner (1997) agrees, but with an important caveat:

> Editors insist – and many believe – that busy readers have no time for the 'rest of the story.' The five *W*s that journalists once revered are often reduced to four. Yes, we can fit the who, what, where and when into the little space on the front page. But skip the why. That fifth *W*, after all, may only raise more questions than it answers.
>
> (Toner 1997: 130)

This is the key issue, in Toner's view. Science journalists who fail to ask the question 'why' are too often providing information that satiates people's curiosity, as opposed to stimulating it. 'Like a weed,' he writes, 'curiosity has a habit of popping up in the wrong place. It can be unruly and hard to control. It is robust and tenacious. And where one question sprouts, many more are bound to follow' (Toner 1997: 128). It is this contagious aspect of curiosity, in his opinion, that renders it 'such a powerful tool for journalism'. Moreover, he contends: 'Readers, listeners, and viewers may appreciate our wit, our incisive grasp of complex issues, and the clarity of our delivery, but by planting the seeds of curiosity, we make our audience accomplices in the pursuit of knowledge' (1997: 129).

If science items can often be, as Petit (1997: 187) suggests, the 'furthest thing from breaking news, this can be their charm'. In discussing his experiences as a reporter interested in the earth sciences, he draws a sharp contrast between the kinds of news stories he typically writes and the more usual kinds of items published alongside them in the same newspaper. 'Like stories in astronomy, or on fossils of prehistoric people,' he writes, 'discoveries about the Earth's history and behavior provide for many people a welcome and invigorating break, a mental escape from the daily diet of human disaster, political skulduggery, and crime news' (Petit 1997: 187). Ropeik (1997), a television news reporter covering science and environmental stories, writes:

> The feedback I've gotten in my 18 years as a journalist leads me to believe that news consumers are curious. They *want* to know. They *want* complicated things explained to them. They have a gee-whiz button waiting to be pushed. I look for ways to push it, in how I organize, how I write, and literally in the tone of voice I use as I narrate my stories.
>
> (Ropeik 1997: 38)

As a result, he places a considerable emphasis in his reporting on being 'simple and clear, given the brevity and mind-numbing nature of television news' (Ropeik 1997: 38). To be effective means playing to television's principal strength, that is, 'to let the pictures do the talking.' The rendering of complex scientific issues into visual images, when done thoughtfully, can spark and sustain the viewer's interest in a way that helps to make core points register. The 'power of pictures to tell the story', he maintains, 'is the television newsperson's greatest tool' (1997: 38).

This issue evidently strikes a resonance with Flatow's (1997) experience in television reporting. In his discussion of the production of long-format science stories for television, he observes:

> The most challenging stories to produce are the ones in the field of science that people 'think' they don't find interesting like physics or chemistry. Just the mention of the words can send viewers to their remote controls. The trick, then, is to disguise the science in the piece, hide it, and spring it on the viewer suddenly. Do this by treating the topic not as a 'science' story but as an Agatha Christie mystery. Your scientists are not white-coated laboratory technicians but science sleuths on the trail of a suspect. Nobody can resist a good whodunit.
>
> (Flatow 1997: 40–1)

The search for such dramatic elements, he concedes, is difficult. 'It may also mean, he adds, 'finding scientists who are good storytellers, personable on camera, and willing to submit to the rigors of television' (Flatow 1997: 41). This last point raises a host of issues for the unsuspecting scientist, as Flatow proceeds to elaborate:

> scientists who agree to become television 'talent' may have no idea of the demands that may be made of their laboratory. Dozens of phone calls interrupt their work. Scripts have to be written and re-written. Schedules must be arranged. Then comes the invasion. Laboratories are besieged by hoards of camera, lighting and sound people. All work stops while those 'TV people' take over. Unsuspecting scientists may balk at the commotion and decide that this is not what they bargained for.
>
> (Flatow 1997: 41)

The logistics involved in finding scientists willing and able to serve as news sources can be formidable. When asked to reflect on how they go about their daily work of identifying those 'newsworthy' sources deserving to be included in a news account, journalists will often claim that they simply follow their 'gut feelings', 'hunches' or 'instincts'. Many insist that they have a 'nose for news', that they can intuitively tell which sources are going to prove significant and which ones are bound to be irrelevant to the news item. It is this issue of how science journalists interact with scientists, then, that is the focus of the next section.

Scientists as sources

The very basis upon which the journalist is able to detect 'news events', as Fishman (1980: 51) points out, rests on the understanding that society is bureaucratically structured. It is this perspective which furnishes the reporter with a 'map of relevant knowers' for newsworthy topics. A science journalist covering a story concerning, say, the possible effects of a nuclear power plant on the health of children in a local community, knows that information officers at the plant, as well as politicians, scientists, nuclear energy lobbyists, health officials, social workers, and environmental groups, among others, will be positioned to offer their viewpoints. 'Whatever the happening,' writes Fishman (1980: 51), 'there are officials and authorities in a structural position to know.' This 'bureaucratic consciousness', to employ his phrase, indicates to journalists precisely where they will have to position themselves to be able to follow the time-line or 'career path' of an event as it passes through a series of interwoven, yet discernible phases. By this rationale, the higher up in this bureaucratic hierarchy the news source is situated, the more authoritative their words will be for reporting purposes. Journalists are predisposed to treat these accounts as factual, according to Fishman (1980: 96), 'because journalists participate in upholding a normative order of authorized knowers in society [and] it is also a position of convenience'. After all, the 'competence' of the source should, by this logic, translate into a 'credible' news story (see also Friedman et al. 1986; Dornan 1990; Gascoigne and Metcalfe 1997; Kiernan 1997; Scanlon et al. 1999).

The phrase 'Get all sides of the story' signifies a vital tenet of journalism that Greenberg (1997), a newspaper editor, emphasizes in his discussion of science reporting. 'Whether a science story involves long-term investigation or quick-turnaround breaking news,' he writes, 'it requires well-rounded, balanced reporting that relies on the savvy and expertise of the writer perhaps more than most other beats' (Greenberg 1997: 97). He proceeds to identify several basic sources for reporting science news stories as follows:

- *Journals*: the steadiest of science news sources, in Greenberg's view, journals encourage a certain form of 'pack reporting' among journalists. The reason, he argues, is straightforward enough: 'If a researcher is going to drop a bombshell, chances are it will land smack in the pages of *Nature*, *Science*, *JAMA*, the *New England Journal of Medicine*, *The Lancet*, or one of scores of other smaller but respected publications' (Greenberg 1997: 97). Publications such as these ones are routinely monitored by the major news organizations, namely because to miss an important research story is 'an acknowledged sin' (see also Young 1997).
- *Meetings*: after journals, scientific meetings are the second likeliest place a scientist will announce a major research finding. 'This does not happen as often as it used to,' Greenberg (1997: 97) argues, 'primarily because refereed journals are considered by many a purer forum in which to divulge advances.' Such meetings are thus less likely to generate 'breaking news', but are nevertheless worth attending by science journalists for 'background' and to 'cultivate sources' for interviews.
- *Breaking news*: just like other journalists, Greenberg points out, science reporters work 'at the whim of events'. Speaking of the situation in Los Angeles, where he is based, he states: 'we're poised to cover an earthquake every day, and to a lesser extent fire and floods. For science writers it is especially important to develop sources expert in such local phenomena that are likely to recur' (1997: 97).
- *Press conferences* or *press releases*: information gathered from these sources needs to be treated with caution. Greenberg argues that a wary eye should be trained on any institution or scientist publicizing their research prior to journal publication, not least because there are 'many real-world agendas that sneak into science, like funding, competition, ambition, and glory' (1997: 97). That said, though, he does concede that

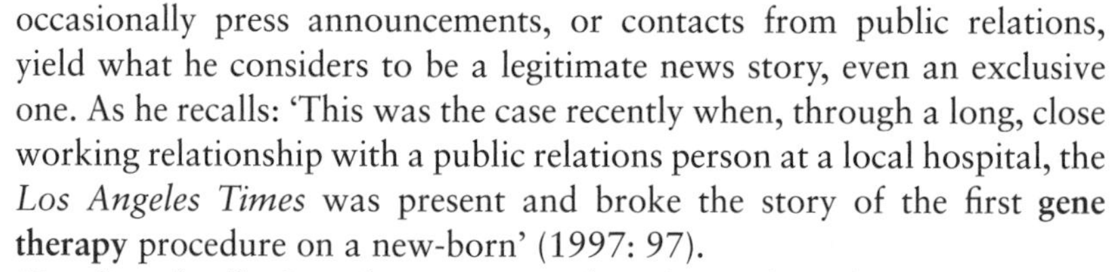

occasionally press announcements, or contacts from public relations, yield what he considers to be a legitimate news story, even an exclusive one. As he recalls: 'This was the case recently when, through a long, close working relationship with a public relations person at a local hospital, the *Los Angeles Times* was present and broke the story of the first **gene therapy** procedure on a new-born' (1997: 97).

- *Unsolicited calls*: Greenberg expresses his advice where this type of source is concerned in blunt terms: 'If "scientists" you don't know want to publish their original "research" in your newspaper or magazine, rather than in a journal, run for cover' (1997: 97–8).
- *Own sources*: a crucial task for any science reporter, indeed for Greenberg possibly the most important part of covering science, is 'building up a cadre of reliable, informed sources that you can call for reaction and comment – both on or off the record – about any range of stories' (1997: 98). Such a task takes time and experience, he argues, as well as 'gaining the trust of such people through consistently accurate and well-written stories' (1997: 98).

Overall, in Greenberg's view, the value of a science news item will ultimately come to rest on the calibre of the sources the reporter has drawn upon to tell the story. In almost every instance, he maintains, the nature of the story determines where the reporter looks for appropriate sources. Such is not always the case, though. 'While the rules for choosing sources on most science stories are clear,' Greenberg (1997: 99–100) writes, 'certain kinds of stories can render those rules as gooey as volcanic mudflows.' Here he cites 'the infamous "cold fusion" experiment' of 1989 to illustrate his point, an occasion where despite the fact that many journalists were deeply sceptical of the claims being made they nevertheless gave the announcement blanket coverage as a major news event (see also Toumey 1996; Bucchi 1998).

The challenge of balancing as many authoritative sources as possible within a news account appears to be a crucial one for the science reporter. Disagreement between sources cannot always be framed in terms of 'right and wrong', instead sometimes it may be more accurately characterized as simply representing varied interpretations. 'Many scientific studies', Young (1997: 116) maintains, 'are so complex, so difficult to do, that their findings do, indeed, lend themselves to two or more interpretations.' The opinion of the research team, he adds, should not be enshrined in truth when others working in the same field may – quite rightly – view the results rather differently. This point is similarly addressed by Harris (1997), an environmental reporter, who argues that in the case of 'extreme' points of view among potential sources, even greater care needs to be taken to maintain journalistic balance. In his view, the kinds of voices heard at the margins of a scientific consensus tend to be either ignored entirely or treated with equal weight in a news story. What is much more appropriate, he maintains, is for the journalist to scrupulously situate each voice within the larger spectrum of opinion so as to enable the news audience to understand their arguments in context. It is important to bear in mind, Harris (1997: 170) points out, that the 'minority view isn't necessarily wrong – just ask Galileo'.

What motivates the scientist to take the journalist seriously? In Rensberger's (1997: 9) experience, they simply 'want the public to know, to understand, and to be on their side in a world too often given to ignorance, fear and superstition'. In discussing how science reporters interact with their sources, Salisbury (1997) argues that the gap between them is usually not as wide as it tends to be in other areas of journalism. Indeed, in his experience, members of the scientific and journalistic communities are in many ways 'natural allies', sharing as they do 'a skeptical approach to information and a devotion to discovering the truth' (Salisbury 1997: 222). Both sides are likely to benefit from what can be a mutually advantageous relationship. Just as news organizations often seek to boost their audience figures by drawing on reports of exciting scientific discoveries, so scientists can attract political and economic support for their research by receiving favourable media treatment. That said, however, this symbiotic relationship can quickly become fraught with difficulties at a number of interrelated levels. Indeed, as Salisbury elaborates:

> Scientists live in a different 'time zone' from reporters. They work on projects for months and years, so a paper that has been out for six months can still seem new to them. Conversely, in some fields it takes six months to a year for a paper to appear in a journal, and by then the scientist has moved onto another topic and considers the paper old news. Because scientists' time frame is so different, they are unlikely to contact the news office at a journalistically appropriate time.
>
> (Salisbury 1997: 220)

In attempting to overview some of the more pronounced ones, Salisbury observes:

- To scientists, the devil is definitely in the details, while journalists are interested primarily in the big picture.
- To scientists, disputation is part of the process of advancing understanding ever closer to truth; to journalists conflict is the source of drama that adds zest to a story.

- Scientists are continually trying to build consensus, while journalists focus on the drama of pro and con.
- To scientists, peer review is an integral part of a process designed to reduce errors. To most journalists, allowing sources to review material before publication is an unacceptable ceding of editorial independence.
- To scientists, technical terms provide added precision and clarity to discourse. To journalists, technical terms constitute a jargon that obfuscates science and makes it incomprehensible to the general reader.

(Salisbury 1997: 222)

Given the significant number of scientists who feel that they have been 'burned' in some way by the media, Salisbury observes that it can be difficult to persuade them that 'the overall benefits to the science community at large are worth the time, energy, and risk involved in dealing with reporters' (1997: 223).

To be effective, then, science journalists need to cultivate a relationship of trust with their sources. Needless to say, the symbiotic nature of the scientist –journalist relationship can quickly come unravelled under certain circumstances, especially where controversial claims are involved. Speaking from his experience as a science reporter, Young (1997) warns of the 'hidden agenda' a scientist can bring to bear in an interview with a journalist:

> The possibility of profiting from their research or a fear of losing their research funding may skew their comments or color their judgement of their work's potential for benefiting humankind. Or they may be wedded too intensely to some cause, such as saving the environment or preventing child abuse, or even to some scientific theory.
>
> (Young 1997: 115)

This cautionary note similarly informs Trafford's (1997) discussion of how health reporters handle their sources. She proceeds to highlight the danger of quoting 'everyone you talk to on an equal basis in the name of "balance" [when] what you're really striving for is fairness and accuracy' (Trafford 1997: 137). As she points out, not every source is equal, nor should it be the case that different viewpoints be reported in equal terms: 'The reader expects you to make the first cut in evaluating the major points in a story' (1997: 137). When it comes to handling potential news sources, then, care needs to be taken to evaluate their relative merits. Public health stories, in her experience as a health editor on a daily newspaper in the US, typically start with government officials:

> In many ways, they [government officials] are like real estate agents: they are often friendly, knowledgeable, and sophisticated. They show you a lot of properties. They want you to be happy and they answer a lot of your questions. But remember, they are always working for the seller – namely, the government, and in some cases, the president, or the governor or major who gave them their job.
>
> (Trafford 1997: 136)

Trafford (1997: 136) discerns several rings of sources flowing outward from this governmental realm of public health officials. The first such ring is composed of major institutions, such as schools of public health, hospital systems, medical schools, and research and policy centres. The second ring encompasses advocacy groups, including disease organizations and grass-roots citizens' groups, as well as lobby groups, which include promotional foundations and trade associations (such as those associated with hospitals or drug companies). In the third ring are what Trafford (1997: 136) calls 'the bystanders', that is, 'those individuals who are more affected by a public health problem and the government's plan to deal with it.' Finally, in the fourth and outermost ring, is the general public.

For those science journalists committed to ensuring that their reporting fulfils this sense of public responsibility, a range of exigent issues emerge. Particularly relevant here is the danger that they will be accused of allowing the 'subjective' opinions of their sources to cloud what should be 'objective' statements of fact. But how can the science reporter establish which facts are 'objective' when their sources disagree? Describing the 'hardships and pit-falls' in science writing, Perlman (1997) observes:

> Journals can be filled with deadly jargon. Claims for statistical significance from randomized double-blind clinical trials can be difficult to challenge. Explaining quarks, subatomic particles with their arcane attributes of strangeness and spin and color and charm, may not gladden the hearts of editors or command the column inches that every science writer knows such stories truly demand. And controversy can arise in every field of science to challenge a reporter's confidence in his or her judgement.
>
> (Perlman 1997: 4–5)

A science editor for a daily newspaper in the US, Perlman is adamant that the science journalist is 'not entitled to bias or conflicts of interest' in their reporting. In the course of making their judgements about what counts as a 'fact', they must eschew hype, and be sure to uphold rules of fairness in handling their sources. That said, however, Perlman (1997: 5) insists that 'they must always recognize that merely counting yeahs and nayes in a scientific controversy fails to serve the public and is rarely a guarantee of fairness.' In his view, good reporting entails more than writing about scientific discoveries or developments in a balanced manner. It also means explaining 'their potential impact and their costs and benefits, even while we present the valid sides of the controversies they generate' (Perlman 1997: 6).

Blum and Knudson (1997: 76), both science reporters, point to 'the continuing culture conflict between scientists and journalists', arguing that it has 'intensified as popular reporting on science has become more extensive and more influential'. The relationship between both groups is slowly evolving, however, as the gap between their respective perceptions arguably narrows. Still, according to Blum and Knudson:

> Many [scientists] remain wary of the media: they don't want to look like show-offs; they aren't certain the audience will understand; and when the reporter ambles in and asks whether CO_2 floats or swims, they aren't sure the reporter gets it either. The whole experience can simply make a researcher nervous, and the result is sometimes obsessive focus on perfecting every detail.
>
> (Blum and Knudson 1997: 76)

This is not to deny, though, that some improvements are being made. Blum and Knudson (1997: 76) suggest that many scientists are beginning to gain a better understanding of the journalistic profession, and in so doing 'may gain a more realistic expectation of media coverage'. Scientists and journalists, they observe, share at least one common goal: 'to make science vivid, real, compelling, and important' (1997: 76).

This is anything but simple in practice, of course. Jarmul (1997: 124), a science journalist, points out how reporters are trained 'to be objective and to keep yourself out of the story' while, at the same time, science teaches them to 'write in an impersonal, passive voice'. Hence his call for science reporters to forget these rules, or at least to re-examine them, in order to become a 'real person' who can 'connect with readers' hearts, not just with their heads' (Jarmul 1997: 124). This need to form points of connection is particularly important where the communication of risk is concerned, the subject of the next section's discussion.

Communicating risk

Each year government agencies from one country to the next publish statistical profiles of risk. In the case of Britain, for example, the Department of

Trade and Industry (DTI) releases an annual surveillance report on domestic accidents. Drawing on hospital statistics, the report routinely shows that the home is a risky place to be. Results for 1999, publicly released in 2001, found that on average 76 people were killed each week in domestic accidents that year – more than died in road accidents (*The Guardian*, 7 June 2001). The top ten causes of accidents in the home were listed as follows:

1. Stairs or steps indoors
2. Carpet or underlay
3. A child causing obstruction or collision
4. Floor surfaces
5. An adult causing obstruction or collision
6. Walking barefoot
7. Concrete surfaces
8. Dogs
9. Outdoor steps
10. Doors

Pensioners and small children were found to be the most vulnerable. 'People can be injured by some quite unlikely items,' commented a DTI spokesperson, 'and more people are injured in the home than anywhere else' (*The Guardian*, 7 June 2001). Evidently each year about 600,000 people require emergency hospital treatment following collisions with other people or objects. Moreover, glossy magazines reportedly cause four times more accidents than chainsaws, while beanbags do more harm than meat cleavers. Over 100,000 accidents a year may be attributed to alcohol consumption, making simple tasks – including removing one's clothes at bedtime apparently – potentially hazardous.

This type of statistical information, used by governments to help develop consumer safety policy, is frequently made the subject of light-hearted commentary in newspapers. An editorial leader in *The Independent* addressing the above DTI report, for example, suggests that life 'is a dangerous business' (7 June 2001). 'Admittedly,' it observes, 'accidents caused by sponges and loofahs are down from 996 to 787, no doubt because of an admirable government initiative, until now sadly ignored. Armchairs, too, have become less aggressive – inflicting a mere 16,000 injuries.' It then proposes that a Royal Commission be established to investigate accidents caused by socks and tights. Humour aside, of course, the difficulties in communicating the realities of risk remain.

A risk of an altogether different nature received extensive news coverage in the United States during the summer of 2001. Several different shark attacks on human swimmers, taking place along the coastlines of Florida,

Virginia and North Carolina, were front-page news across the country. Despite being isolated incidents, many news reports alerted members of the public to the 'dramatically increasing risks' of swimming in the ocean, typically making vivid references to 'a spate of shark attacks' or 'a wave of shark attacks'. As a result, to the extent that these separate incidents were held to be indicative of the 'Season of the Shark' (a *Time* magazine cover story headline), such reports encouraged the perception that there was an underlying pattern or trend emerging. If for some journalists there was something more than coincidence at work, however, it was 'the dull duty of science' – to use a phrase from a *New York Times* editorial – to point out that these incidents were 'grouped together only in time and by the fact that every victim was in the water when the attack took place' (9 September 2001). The attendant risk calculation, it follows, must be placed in context. Indeed, as the editorial goes on to point out, research from the Statistical Assessment Service maintains that 28 children were killed by falling television sets in the US between 1990 and 1997, four times as many people who were killed by great white shark attacks in the twentieth century. 'Loosely speaking,' the editorial observes, this research suggests that 'watching "Jaws" on TV is more dangerous than swimming in the Pacific.'

While the media frenzy over the risk of shark attacks was unwarranted, the problem of how best to define what counts as a 'risk' – as opposed to, say, a 'threat' or a 'hazard' – remains an awkward one. The subject of considerable debate both within and beyond scientific communities, meanings of the term often varying quite dramatically from one user to the next. In its technical sense, however, risk is usually defined as the calculated probability of an adverse consequence (such as a danger, harm or loss) arising because of a specific action or process. Adams (1999: 285) identifies three broad categories of risk:

- directly perceptible risks: e.g. traffic to and from landfill sites
- risks perceptible with the help of science: e.g. cholera and toxins in landfill sites
- virtual risks – scientists don't know/cannot agree: e.g. BSE [bovine spongiform encephalopathy or 'mad cow disease' or CJD Creutzfeldt–Jakob disease] and suspected carcinogens.

Self-described 'risk managers' tend to focus on the first two of these three categories. The reason, Adams (1999: 285) argues, is that '[q]uantified risk assessments require that the probabilities associated with particular events be known or be capable of plausible estimation'. Scientists, as many of them are quick to acknowledge, tend to frame issues of risk in terms of probabilities which are little more than confident expressions of uncertainty. 'When scientists cannot agree on the odds,' writes Adams (1999: 285), 'or the underlying causal mechanisms, of illness, injury or environmental harm, people are liberated to argue from belief and conviction' (see also Friedman et al. 1999; Scanlon et al. 1999).

Scientists' perceptions of risk, one study after the next suggests, can be at serious odds with those held by members of the public. Research commissioned on public perceptions of risk by Britain's Parliamentary Office of Science and Technology (POST 1996), for example, provides a series of pertinent insights. In addition to the actual size of the risk, a variety of different factors are identified which appear to influence public perceptions:

- *Control*: people are more willing to accept risks they impose upon themselves, or they consider to be 'natural', than to have risks imposed upon them.
- *Dread and scale of impact*: fear is greatest where the consequences of a risk are likely to be catastrophic rather than spread over time.
- *Familiarity*: people appear more willing to accept risks that are familiar rather than new risks.
- *Timing*: risks seem to be more acceptable if the consequences are immediate or short term, rather than if they are delayed – especially if they might affect future generations.
- *Social amplification and attenuation*: concern can be increased because of media coverage or graphic depiction of events, or reduced by economic hardship.
- *Trust*: a key factor is how far the public trusts regulators, policy-makers, or industry. If these bodies are open and accountable – being honest, admitting mistakes and limitations and taking account of differing views without disregarding them as emotive or irrational – then the public is more likely to place credibility in them.

(POST 1996)

These factors, taken together, contribute to a better understanding of why some risks are perceived as being more serious than others. Each in turn highlights, to varying degrees, the significance of the media in shaping these perceptions. That is to say, confronted by scientific uncertainty where risks are concerned, 'ordinary' or 'lay' members of the public are likely to turn to the news media, in particular, for a greater understanding of what is at stake. Journalists are charged with the responsibility of imposing meaning upon uncertainties, that is, it is expected that they will render intelligible the underlying significance of uncertainties for their audiences' everyday experiences of modern life. More often than not, news accounts will offer the assurance that a potential risk will remain uncertain only until further research and scientific investigation are able to provide the expected clarity and certitude (see also Adam 2000).

Risk, as Hornig (1993) points out, thus becomes a deeply political issue. 'In a post-industrial democracy,' she writes, 'confronting the social acceptability of the risks of emerging technologies is an everyday form of crisis' (Hornig 1993: 95). In seeking to investigate the decisive influence of the media in shaping public debate over risk policy issues, she draws attention to the ways in which diverse political actors use scientific opinion to advance their interests. Specifically, she identifies two distinct approaches to the evaluation of risks, namely the 'rationalist' and the 'subjectivist' positions. Both of these perspectives incorporate – either explicitly or implicitly – an opposing range of assumptions about scientific opinion in relation to how different levels of risk are best determined.

The rationalist position, Hornig (1993: 96) maintains, 'holds that it is theoretically possible, if sufficient data could be collected and various technical problems of analysis solved, to arrive at an absolute measure of the riskiness associated with any technological innovation'. For advocates of this position, this theoretical measure may then serve, in turn, as a yardstick of sorts against which the sway of public opinion and media representations may be calculated. At issue, in their view, is the need to improve risk-related decision-making by ensuring that systematic 'distortions' are recognized as such and set right through public education. As a result, Hornig (1993: 96) observes that from this position 'media accounts of risk are typically judged on how accurately they reflect the scientific point of view and how well they contribute to public education designed to eradicate wrong thinking'.

Sharply counterpoised against this approach to risk is the 'subjectivist' position. Here Hornig acknowledges the contribution made by various psychometric studies of risk perception to the development of this position over the past few years. Of particular import is the line of argument which holds that 'the evaluation of risk information takes place in a social context and involves value judgements and priorities – that is, that this process is inherently subjective' (1993: 96). Risk, it follows, is not an objective condition somehow existing outside of human observation or interpretation. Rather, research undertaken from this perspective focuses primarily on the social process in and through which particular definitions of risk are constructed. That is to say, under scrutiny are these contending definitions precisely as they are being mobilized to lend credence to a particular understanding of the risk issue in question. 'The media figure prominently in this picture', Hornig (1993: 96) notes, 'as they vocalize and therefore legitimize some points of view (often those of established institutional news sources) and ignore others.' This often subtle process of legitimizing certain definitions of risk at the expense of alternative ones serves to illuminate the clash of competing institutional interests. 'News accounts', she argues, 'inevitably contribute to definitions-of-the-situation that serve some interests in preference to others, the journalistic ethic of objectivity (which echoes in interesting ways the rationalist notion of risk) notwithstanding' (1993: 96).

It is hardly surprising, of course, that news coverage of risk issues is found wanting from the vantage point of both positions, respectively. If, from the rationalist perspective, news accounts are frequently blamed for distorting and politicizing 'technical' risk issues, Hornig maintains, from a subjectivist perspective they are often blamed for over-representing the scientific point of view in positive terms. Risk reporting, she argues, 'seems destined to please no-one – a problem that is much deeper and more difficult than a question of whether scientific and environmental journalists get their technical facts straight, or whether media accounts are pro- or anti-technology' (1993: 97). Of particular concern, then, is the extent to which the media help to set the agenda of public concerns, especially with regard to defining or 'framing' risk issues for their audiences.

In the case of her own research into public responses to media accounts of technological risk, Hornig accepts the subjectivist premise that levels of risk are contingent upon a given interpretive context. In dispensing with any idea of attempting to 'measure' the relative 'degree of divergence from the "correct" interpretations' of risk available, she seeks to study instead the arguments that underlie actual 'lay' or 'ordinary' people's risk evaluations. Lay publics, her focus group research suggests, work with an 'expanded vocabulary of risk' that 'takes into account a broader and in a sense more sophisticated range of factors than do rationalist measures of risk' (1993: 98). Such factors include, for example, a 'broad variety of considerations related to technologies will fit into and be controlled within the social system: the technologies' compatibility with ethical principles, procedures for their regulation, where potential or alleged risks are undertaken knowingly or unknowingly, and so on' (1993: 106). Thus to examine how members of the public negotiate the significance of the various risks indicative of modern society, researchers will need to extend the scope of their investigations far beyond the limits of rationalist evaluative frameworks, and their attendant cost–benefit analyses (see also Tulloch and Lupton 2001).

Critical lines of inquiry, it follows, have to recognize the importance of accounting for the media's construction of risk as part of the lived politics of the everyday. In this context, the work of Ulrich Beck (1992a, 1995, 2000) has proven to be highly influential. Particularly consequential, for example, is his conception of the 'relations of definition' underpinning media discourses which condition what can and should be said about risks, threats and hazards by 'experts' and 'counter-experts', as well as by members of the lay public. The analysis of risk, he argues, needs to account for the media's structurating significance in the formation of public opinions about risk. To clarify, Beck (1995) accords to the media a crucial role in the organization and dissemination of knowledge about economic decision-making and political control vis-à-vis the uncertainties associated with risks:

> The system of institutionally heightened expectations forms the social background in front of which – under the close scrutiny of the mass media and the murmurs of the tensely attentive public – the institutions of industrial society present the dance of the veiling of hazards. The hazards, which are not merely projected onto the world stage, but really threaten, are illuminated under the mass media spotlight.
>
> (Beck 1995: 101)

Important questions therefore arise as to who in the media wields this spotlight, under what circumstances and, moreover, where it is (and is not) directed and why. It follows that it is of the utmost significance how issues of proof, accountability and compensation are represented in and by media discourses. Risks, as Beck (1992a: 23, original emphasis) writes, can 'be changed, magnified, dramatized or minimized within knowledge, and to that extent they are particularly *open to social definition and construction*'. The ways in which journalists help to mediate the limits of risk, as always both in conjunction with and opposition to other institutions across society, consequently need to be carefully unravelled for purposes of analysis.

This process of mediation involves a series of procedures for knowing the world and, equally importantly, for not knowing that world as well. It is fraught with uncertainty, ambiguity and contradiction. As noted above, the preferred 'models' of the scientist, for example, do not 'translate' easily into the reportorial strategies of the journalist anxious to convey their meaning to the intended audience. 'Risk societies', writes Beck (1998: 19), 'are currently trapped in a vocabulary that lends itself to an interrogation of the risks and hazards through the relations of definition of simple, classic, first modernity.' As a result, he continues, this vocabulary is 'singularly inappropriate not only for modern catastrophes, but also for the challenges of manufactured uncertainties' (Beck 1998: 19). In the case of scientific perspectives, then, they must therefore undergo a process of journalistic narrativization before they are likely to 'make sense' to a public facing unknown and barely calculable risks. 'Dispassionate facts' must be marshalled into a 'balanced' news story, ideally one with a distinct beginning, middle and end, as well as with easily identifiable 'good' versus 'evil' conflicts. This struggle to narrativize the scientific world necessarily situates journalists at the point where, as Beck (1992b) observes, the antagonisms between those who produce risk definitions and those who consume them are at their most apparent. Daily newspaper reading, as he notes, becomes 'an exercise in technology critique'.

This is to suggest, then, that the identification of the slips, fissures, silences and gaps in media reporting needs to be simultaneously accompanied by a search for alternatives. New ways need to be found to enhance the forms and practices of science journalism in a manner consistent with today's moral and ethical responsibilities for tomorrow. Many of the salient issues discussed in this chapter relate to the 'monitoring' and 'surveillance' functions Beck attributes to the media as privileged sites for larger definitional struggles over the scale, degree and urgency of what are incalculable risks. In seeking to denaturalize the ways in which the media process certain voices as being self-evidently 'expert' or 'authoritative' while simultaneously framing others as lacking 'credibility', it is this very self-evidentness which needs to be recognized as a terrain of discursive struggle. 'In the case of risk conflicts', Beck (1998: 15) declares, 'bureaucracies are suddenly unmasked and the alarmed public becomes aware of what they really are: *forms of organized irresponsibility*.' Media institutions are very much implicated in these 'forms of organized irresponsibility'; their representations of the ideological contests being waged over the right to characterize the consequences of risks are typically anything but 'impartial' or 'objective'. Regardless of what some journalists might insist, facts do not 'speak for themselves', and voices calling into question scientific rationalities are not, by definition, 'irrational', 'misinformed' or 'anti-science' (see also Cottle 1998; Macnaghten and Urry 1998; Adam et al. 2000).

The next chapter's discussion of media representations of 'the environment' will return to the writings of Beck on the 'risk society' in the first instance, before proceeding to offer an evaluative appraisal of the diverse ways in which journalists frame the realities of environmental risks, threats and hazards. A key point of departure, then, is this recognition that the taken-for-granted, seemingly common sensical assumptions underpinning the media's preferred relations of definition are pivotal to the way we as members of the public understand, negotiate and challenge the uncertainties of today's risk society. Indeed, as Beck (2000: xiv) observes, this society 'can be grasped theoretically, empirically and politically only if one starts from the premise that it is always a knowledge, media and information society at the same time – or, often enough as well, a society of non-knowledge and disinformation'.

Further reading

Adam, B., Beck, U. and van Loon, J. (eds) (2000) *The Risk Society and Beyond: Critical Issues for Social Theory*. London: Sage.

Allan, S. (1999) *News Culture*. Buckingham and Philadelphia, PA: Open University Press.

Beck, U. (1992) *Risk Society: Towards a New Modernity*. London: Sage.

Blum, D. and Knudson, M. (eds) (1997) *A Field Guide for Science Writers*. New York: Oxford University Press.

Bucchi, M. (1998) *Science and the Media: Alternative Routes in Scientific Communication*. London: Routledge.

Friedman, S.M., Dunwoody, S. and Rogers, C.L. (eds) (1986) *Scientists and Journalists*. Washington, DC: AAAS.

Friedman, S.M., Dunwoody, S. and Rogers, C.L. (eds) (1999) *Communicating Uncertainty*. Mahwah, NJ: Lawrence Erlbaum.

Gregory, J. and Miller, S. (1998) *Science in Public*. Cambridge: Perseus.

Hargreaves, I. and Ferguson, G. (2000) *Who's Misunderstanding Whom? Bridging the Gulf of Understanding Between the Public, the Media and Science*. London: ESRC / British Academy.

Nelkin, D. (1995) *Selling Science: How the Press Covers Science and Technology*, 2nd edn. New York: W.H. Freeman.

Scanlon, E., Whitelegg, E. and Yates, S. (eds) (1999) *Communicating Science, Reader 2*. London: Routledge.

MEDIA, RISK AND THE ENVIRONMENT

> Environmental degradation and hazards pose threats that differ from the calculable risks associated with car accidents, thefts and house fires, in that they are neither just unintended consequences of rational actions, nor mere side effects, but *endemic* to the scientific innovations and economic practices characteristic of the industrial way of life: for hazards arising from the industrial way of life, the past gives no guidance for the future, provides no basis upon which to calculate and quantify risk.
>
> (Barbara Adam, social theorist)

> . . . there is a big difference between those who take risks and those who are victimised by risks others take.
>
> (Ulrich Beck, social theorist)

Public debates over how to best sustain 'the environment' have arguably never been more openly contested than they are today.[1] Configurations of the environment as a purely natural realm, one existing outside of the lived dynamics of human activity, are increasingly being subjected to challenges from across the media spectrum. Recurrently called into question are 'traditional' or 'modernist' conceptions of the 'natural' as being distinct from the 'human', as if it was reducible to the observable 'raw materials' of the world. Indeed, at a time when growing numbers of public commentators are engaged in disputes over the potential dissolution of the 'actual' into the 'virtual', some argue that the language of environmentalism is fast becoming anachronistic in its appeal to a singular 'nature' to sustain its convictions.

Soper (1995), in her book *What is Nature?*, aptly describes a range of characteristics typically projected upon 'nature'. In the case of popular imagery, she observes:

> Nature is both machine and organism, passive matter and vitalist agency. It is represented as both savage and noble, polluted and wholesome, lewd and innocent, carnal and pure, chaotic and ordered. Conceived as a feminine principle, nature is equally lover, mother and virago: a source of sensual delight, a nurturing bosom, a site of treacherous and vindictive forces bent on retribution for her human violation. Sublime and pastoral, indifferent to human purposes and willing servant of them, nature awes as she consoles, strikes terror as she pacifies, presents herself as both the best of friends and the worst of foes.
>
> (Soper 1995: 71)

Different ways of talking about nature accentuate the contradictory meanings its usage engenders, particularly where discourses of risk are concerned. Nature, as Wilson (1992: 12) writes, 'is filmed, pictured, written, and talked about everywhere', leading him to similarly maintain that there are in fact many different natures being rearticulated via media discourses. Here, though, he makes the crucial point that the current crisis around 'the natural' is not only 'out there in the environment'. Rather, he argues, it is also a 'crisis of culture', one which 'suffuses our households, our conversations, our economies' (Wilson 1992: 12; see also Macnaghten and Urry 1998; Franklin et al. 2000; Lindahl Elliot 2001).

In this context, Adams (1999) usefully outlines a typology of 'myths of nature'. These myths, he argues, describe 'various preconceptions about nature that inform risk-taking decisions in such circumstances' (Adams 1999: 295). Each of the myths corresponds, in turn, with a distinctive risk-management style:

- *Nature benign*: nature, according to this myth, is predictable, beautiful, robust, stable, and forgiving of any insults humankind might inflict upon it [. . .] Nature is the benign arena of human activity, not something that needs to be managed. The [risk] management style associated with this myth is therefore relaxed, exploitative, laissez-faire.
- *Nature ephemeral*: here, nature is fragile, precarious and unforgiving. It is in danger of being provoked by human greed or carelessness into catastrophic collapse. The objective of [risk] management is the protection of nature from Man. People, the myth insists, must tread lightly on the earth. The guiding management rule is the **precautionary principle.**
- *Nature perverse / tolerant*: this is a combination of modified versions of the first two myths. Within limits nature can be relied upon to behave predictably. It is forgiving of modest shocks to the system, but

> care must be taken [. . .] Regulation is required to prevent major excesses, while leaving the system to look after itself in minor matters. This is the ecologist's equivalent of a mixed-economy model. The [risk] manager's style is interventionist.
>
> - *Nature capricious*: nature is unpredictable. The appropriate [risk] management strategy is again laissez-faire, in the sense that there is no point in management. Where adherents to the myth of a nature benign trust nature to be kind and generous the believer in nature capricious is agnostic; the future may turn out well or badly, but in any event, it is beyond his [or her] control. The non-manager's motto is *que sera sera*.
>
> (Adams 1999: 295–6)

Precisely which approach to environmental risk management is judged to be the most 'appropriate' will thus depend, in part, upon which 'myth of nature' is accepted as being 'natural'. Further 'myths' to these ones may be envisaged as well, of course, but of particular significance for our purposes are the ways in which these ideological tensions around 'nature' and 'the environment' are negotiated across the field of the media in terms of risk.

Accordingly, to further elucidate the means by which the media help to shape public perceptions of environmental risks, we return to Beck's (1992a, 2000) conception of the 'risk society' in the next section. The 'public eye of media', as he points out, 'takes on a key significance in the risk society' (Beck 2000: xiii). That is to say, news media institutions 'play a decisive role [with] their portrayal of conflicting definitions of risk, that is, their representation, or construction, of risks and uncertainties' (2000: xiii). As will be shown, this conceptual intervention brings to the fore a number of consequential issues associated with the cultural politics of environmental risks which otherwise tend to be obscured in related types of inquiries.

Living in the risk society

'Today,' writes Beck (1998: 10–11), 'if we talk about nature we talk about culture and if we talk about culture we talk about nature.' As he proceeds to elaborate, any conception of modern society which is sensitive to issues of risk needs to recognize that the world is much more open and contingent than is typically postulated via the theoretical distinction nature/culture:

> When we think of global warming, the hole in the **ozone layer**, pollution or food scares, nature is inescapably contaminated by human activity. This common danger has a levelling effect that whittles away

> some of the carefully erected boundaries between classes, nations, humans and the rest of nature, between creators of culture and creatures of instinct, or to use an earlier distinction, between beings with and those without a soul [. . .] In risk society, modern society becomes reflexive, that is, becomes both an issue and a problem for itself.
>
> (Beck 1998: 11)

To clarify, Beck designates contemporary societies as risk societies, social formations where large-scale environmental risks, dangers and hazards have become a central structuring feature (just as the logic of wealth creation and distribution dominated industrial society; see also Hannigan 1995; Goldblatt 1996; Rutherford 1999). The known threat of catastrophes, it follows, is always everywhere present. 'The category of risk', Beck (2000: xii) writes, 'becomes central at the point where the apparent boundaries of nature and tradition dissolve into decisions.' Processes of environmental degradation, it follows, cease to be entirely calculable or predictable; their consequences defy definition in relation to fixed notions of time, space or place.

To live in the risk society, Beck suggests, is to live in the 'hazardous age of creeping catastrophe', where social debates are increasingly dominated by the 'dark sides of progress' (see also Adam 1998; Adam et al. 2000; Irwin 2001). The riskiness of everyday life is becoming evermore apparent as uncertainties deepen. Just as people become more reliant on 'experts' to manage and control risks, their trust in these same experts continues to weaken. Paradoxically, the risk society produces and seeks to legitimize the very hazards which are beyond the control of its institutions. As Beck (1996a) explains:

> The entry into risk society occurs at the moment when the hazards which are now decided and consequently produced by society *undermine and/or cancel the established safety systems of the provident state's existing risk calculations*. In contrast to early industrial risks, nuclear, chemical, ecological and genetic engineering risks a) can be limited in terms of neither time nor place, b) are not accountable according to the established rules of causality, blame and liability, and c) cannot be compensated or insured against. Or, to express it by reference to a single example: the injured of Chernobyl are today, years after the catastrophe, not even all *born* yet.
>
> (Beck 1996: 31, original emphasis)

The indeterminacies judged to be characteristic of environmental hazards are far from temporary; instead they are necessarily contingent and indeterminate (see Lash et al. 1996; Robertson et al. 1996; Macnaghten and Urry

1998; Irwin 2001). Closely linked, then, is Beck's suggestion that through the threat to people, animals, plants and the elements that sustain life, modern societies such as those in the Western world are re-experiencing their interdependence and oneness with nature (see also Lindahl Elliot 2001). They are encountering forms of knowledge that have been progressively eroding the organizing tenets of earlier 'metanarratives' held to be consistent with the 'traditions' of their intellectual heritage.

The driving forces of global modernization impress upon many of us living in modern societies the realization of interconnectedness. Levels of what Beck (1992a) calls a 'human consciousness of nature' are both 'wounded and awakened' in the awareness of hazards:

> people have the experience that they breathe like the plants, and live *from* water as the fish live *in* water. The toxic threat makes them sense that they participate with their bodies in things – 'a metabolic process with consciousness and morality' – and consequently, that they can be eroded like the stones and the trees in the acid rain. A community among Earth, plant, animal and human being becomes visible, a *solidarity of living things*, that affects everyone and everything equally in the threat.
>
> (Beck 1992a: 74, original emphasis)

This dissolution of the culture/nature dichotomy is the historical product of industrial society's activities which impose on an earth community of living and inanimate beings the unintended consequences of environmental risks. As Beck (1992b) suggests with regard to examples such as nuclear or chemical contamination, 'we experience "the end of the other", the end of all our carefully cultivated opportunities for distancing ourselves and retreating behind this category' (Beck 1992b: 109). The 'social explosiveness of hazard' brings about in its wake the implosion of industrial notions of risk perception.

When examining environmental risks it is immediately apparent that there are no times, spaces or places outside of 'nature', just as there are no positions from which the journalist may 'objectively' observe. Everyone is implicated, if in exceedingly uneven positions of power. Nature, as Beck (1992a: 80, original emphasis) suggests, 'is *neither* given *nor* ascribed'. Nevertheless, one media-contested dispute between 'experts' after another is more than likely to reaffirm as 'common sense' the assumption that scientific knowledge must be 'harnessed' so that nature can be 'controlled' or 'managed' in the interest of promoting economic expansion and capital accumulation. This assumption has become *normalized*, to varying degrees, as a largely unquestioned feature of 'media debate', thereby foreclosing potential options and choices for engaging with the conditions Beck (1992a) deems to be characteristic of the 'ecological field of conflict'. Here I hasten to add, however, it is not my intention to suggest that there is a 'reality out there' that is being falsified (consciously or not) by journalists or other media practitioners. What I do want to argue, however, is that environmental knowledge is always mediated in and through contending discourses organized to advance certain truth-claims over and above alternative ones. This capacity to define potential risks and hazards is broadly aligned with the distribution of power among 'credible', 'authoritative' and 'legitimate' definers of 'reality' across the media field.

Far from simply 'reflecting' the reality of environmental risks, it follows, media discourses effectively provide contingently codified (rule-bound) definitions of what should count as the reality of environmental risks. This constant, always dynamic process of mediation is accomplished primarily in ideological terms, but not simply at the level of the media text *per se*. Instead, the fluidly complex conditions under which these texts are both produced and consumed or 'read' will need to be accounted for via critical modes of inquiry. As Beck (1995) writes:

> All this gives the mass media a leading role in sounding the social alarm – so long as they dispose of the institutionally guaranteed right to select their own topics. What eludes sensory perception becomes socially available to 'experience' in media practices and reports. Pictures of tree skeletons, worm-infested fish, dead seals (whose living images have been engraved on human hearts) condense and concretize what is otherwise ungraspable in everyday life.
>
> (Beck 1995: 100)

Indeed, it follows that in modern societies it is virtually impossible to even begin to think through the following 'relations of definition' identified by Beck (1998, 2000) without recognizing the centrality of the media to these processes:

1 Who is to determine the harmfulness of products or the danger of risks? Is the responsibility with those who generate those risks, with those who benefit from them, or with public agencies?
2 What kind of knowledge or non-knowledge about the causes, dimensions, actors, etc., is involved? To whom does that 'proof' have to be submitted?
3 What is to count as sufficient proof in a world in which we necessarily deal with contested knowledge and probabilities?
4 If there are dangers and damages, who is to decide on compensation

> for the afflicted and on appropriate forms of future control and regulation?
>
> (Beck 1998: 18)

Underlying these and related types of questions is the recognition that we are inescapably dependent on the media to comprehend the 'world out there' beyond our immediate experience. It is these complex processes of mediation which work to turn environmental risks into a 'reality' to be represented on the pages of our newspapers or on our television screens. Consequently, then, we have to disrupt or 'denormalize' the very familiarity of journalistic conventions, especially where they are inscribed in accordance with 'common sense' commitments to privileging scientific rationality as an adequate foundation of knowledge in and by itself. 'The consequence for politics', as Beck (1992a: 197–8) points out, 'is that reports on discoveries of toxins in refuse dumps, if catapulted overnight into the headlines, change the political agenda. The established public opinion that the forests are dying compels new priorities' (see also Cottle 1998; McGuigan 1999; Tulloch and Lupton 2001).

Beginning in the next section, the emergence and ensuing development of a number of analytical approaches committed to examining how the media portray the environment as a social problem is highlighted. In appraising the formative influence of these early investigations, I attempt to show that much of this work succeeded in securing a range of vital insights into how media accounts construct certain preferred definitions of environmental realities.

Mapping 'the environment'

It is something of a truism for many researchers interested in the circumstances surrounding the emergence of public discourses of 'the environment' that everything changed in 1969. That was the year startling images of planet Earth were relayed from the surface of the moon, the impact of which – many have maintained since – fundamentally recast the environmental perceptions of what was for a fleeting instant a near-global citizenry. British environmental journalist, North (1998), recalls this event with a certain amount of passion:

> Remember that image of the planet earth floating alone in the universe? The US astronauts beamed it back to us. That's the image that spoiled it all. That's when we started talking nonsense about the world. Suddenly the world's happy materialists, and its happy consumers, were turned into guilt-ridden 'greens'. They saw the spaceman's view of the planet

> and they thought they saw something which was a fragile, static set of natural communities. Actually, nature is, of course, robust, and it's in constant tension. It's dynamic and it is absolutely full of opportunism [. . .] What's more, nature's very nasty and it's extremely violent.
>
> (North 1998: 85)

These images evidently contributed to what may be appropriately described as an 'epistemological break' at the level of media representation. Never again would claims about the relative effects of human societies on 'the natural world' fit quite so comfortably into 'traditional' categories.

'By conquering the frontier of outer space,' Schoenfeld et al. (1979: 43) write of the response in the US, 'Americans seemingly discovered another frontier, the search for a state of harmony between humankind and the only earth we have; and reporters and editors watched – and responded.' The enhancement of media attention directed to conservation issues around this time (*Time* magazine, for example, introduced an 'Environment' section in its 1 August 1969 edition) was also attributable to other events. Included here would be the disastrous break-up of a super-tanker off the coast of Cornwall, England in 1967 and, in the US, the 1969 Santa Barbara Channel-Union Oil leak. These events engendered sustained debate among interested stakeholders, such as government agencies, industry spokespeople, scientists, citizen-action pressure groups, consumer organizations and academics, thereby ensuring that a range of what were often acrimonious disputes featured prominently on the news agenda. That said, however, it quickly became apparent to many reporters seeking to translate the complex language routinely employed by these stakeholders that a new vocabulary would be required. What was needed were ways to interpret the environment as 'news' for the benefit of audiences anxious to understand the long-term implications of these events for their own lives.

Critical examinations of the pertinent types of news coverage produced during this period indicate that the preferred terms of 'conservation' were gradually being supplanted by new, or at least sharply redefined, ecological concepts explicitly associated with 'the environment' as a social problem. Emergent forms of environmental discourse typically stretched notions of conservation beyond the earlier emphasis on natural resources. Regarded as important was the need to encompass the human species as an organism in need of protection in the face of possible extinction (see Schoenfeld 1980; Lowe and Morrison 1984; Hannigan 1995; Neuzil and Kovarik 1996; Anderson 1997, 2000). This shift in the rhetorical strategies of stakeholders posed an acute challenge to the seemingly 'common sensical' division between 'nature' and 'humanity' that had been a recurrent – if largely tacit –

feature of news reportage and, as a result, much public debate. More specifically, it was becoming increasingly evident from one news organization to the next that to the extent a 'classic environmental story' could be identified, it was in practice more accurately described – to quote one reporter – as a 'business-medical-scientific-economic-political-social-pollution story' (cited in Schoenfeld et al. 1979: 52). This growing awareness that 'the environment' defied institutional routinization vis-à-vis existing news beats (in contrast with, for example, the courts, city hall, health, education, sports, finance and so forth) led to the invention and widespread implementation of 'environment beats' in order to generate the right types of 'newsworthy' items.

The ensuing 'information explosion' about 'the environment' underpinned its transformation into a 'hot news story'. In the US, according to Sachsman (1976: 54), events like the Santa Barbara oil spill had 'caused print and broadcast editors to begin taking seriously their own local problems of air and water pollution, overcrowding, and the loss of natural resources' (see also Molotch and Lester 1974). Significantly, however, the steady rise in environmental awareness during the 1970s in countries like those in North America (Earth Day 1970, for instance, received extended news coverage in the US) and Europe sparked a corresponding intensification of efforts among corporate public relations (PR) agencies. If throughout the 1960s the consequences of industrial pollution, to take one example, were widely seen to be an inevitable price to be paid for enjoying the benefits of modern society, this was largely attributable to the concerted promotional campaigns mobilized by PR practitioners (see also Wilson 1992; Beder 1997). 'In 1961,' according to Powell and Leiss (1997: 231), 'the US federal government had 1,164 people working as writers/editors and public affairs specialists. By 1990 the number in public information jobs was nearly five thousand, making the federal government the nation's largest single employer of public information officers'. As they proceed to point out, those 'sources that are best organized to provide technical information to journalists in an efficiently packaged form have a great deal of control over what ultimately appears as news' (Powell and Leiss 1997: 231). At stake, in the view of these advocates, was the need to regulate the parameters of public debate to advantage, an objective which became increasingly difficult in the 1970s as journalists began to look further afield for their sources of information. Also during this period, and in conjunction with the ascent of environmental beats or bureaux, news organizations were recognizing the need to hire specialist personnel who were better able to critically appraise the scientific and technical claims being made by special interest groups (see also Miller and Riechert 2000).

Nevertheless, according to many commentators at the time, as the mid-1970s approached there occurred a marked decline in the level of public interest in environmental issues. Investigating the situation in Canada, for example, Parlour and Schatzow (1978) contend that an explanation for this decline can be linked to the following factors:

1 The decline in interest was initiated by the mass media through the sudden displacement of environmental issues by other issues, such as energy, unemployment and inflation.
2 The surficial impact of mass media coverage of environmental issues on public attitudes and behaviour led to a very rapid decline in public concern once the reinforcing and sustaining influence of the mass media disappeared.
3 The institutionalisation of concern within the political system after 1970 undoubtedly led to the feeling that the problems were being 'solved'; not an unexpected response given the ease with which the public were able to justify the transfer of responsibility for these problems from the individual to the institutional level.

(Parlour and Schatzow 1978: 15)

Reaffirmed throughout their study of the media's environmental coverage from 1960 up to 1972 (the year of the UN Conference on the Environment) is the assumption that public awareness and concern for environmental issues correlates with the relative amount of coverage being generated by the news organizations. Moreover, media attention is held to be a key determinant in the legitimization of the environment as a major political issue, one in need of continuous monitoring by the state at a structural level. In the US, for example, ecological concerns were institutionalized in government agencies such as the Environmental Protection Agency and the Council on Environmental Quality in the 1970s, both of which were regularly treated by journalists as authoritative news sources. Perhaps most significant of all, however, is Parlour and Schatzow's (1978) conviction that the institutionalization of environmental concerns by the state virtually ensures that they will be ultimately rendered subordinate to the priorities of economic growth and performance. News organizations, as one study after the next suggests, are inclined to do very little to challenge this outcome (see also Howenstine 1987; Linné 1991; Hansen 1993; Neuzil and Kovarik 1996).

A range of critical analyses of how the media routinely engaged with different environmental issues, events and crises during the latter part of the 1970s and into 1980s provide further evidence to support this line of argument. Evidently it is only by the start of the latter decade that something resembling an 'ecological conscience' finally began to penetrate the typical newsroom in countries such as Britain and the US in a serious way. Drawing on data collected via his interviews with environmental reporters in the US, Schoenfeld (1980) cites the words of one of them which neatly pinpoints a crucial question at the heart of environmental journalism at the time:

> Do you give readers what they should know or something they will read? The challenge of the environmental beat is to convey a sense of immediacy and pertinence, usually by telling the story in human terms [. . .] I try to find the human element while writing about an increasingly complex world of bewildering facts and figures. Every beat needs that, but this beat demands it.
>
> (cited in Schoenfeld 1980: 462)

This prioritization of the 'human element', and with it the corresponding emphasis on the *extraordinary* at the expense of the *ordinary*, decisively shapes what gets reported and how. A focal point for several research inquiries during this period concerned the ways in which news coverage typically places a priority on spectacular (and, in the case of television, visually sensational) events. Reporting about natural disasters, such as earthquakes, hurricanes, drought or floods (see Adams 1986; Gaddy and Tanjong 1986; Sood et al. 1987; Krug 1993; Simon 1996; Major and Atwood 1997), is shown to be consistently preferred by news organizations over and above 'everyday' hazards. Included among the latter are pesticide-dependent farming, exposure to asbestos dust, lead in petrol or sunbathing. Greenberg et al.'s (1989) analysis of US network television reporting on environmental risk in the mid-1980s effectively summed up these tendencies. Specifically, their research found that 'the disproportionate coverage – from the scientific perspective on risk – of chemical incidents, earthquakes, and airplane accidents probably reinforces the public's well-documented tendency to overestimate sudden and violent risks and underestimate chronic ones' (Greenberg et al. 1989a: 276).

Certainly studies of environmental reporting over recent years have discerned an array of environmental concerns which did not receive sustained high-level news coverage despite being long-term threats, in part due to their non-event oriented characteristics. That is to say, these lines of inquiry have pinpointed the extent to which this coverage is event-centred as opposed to issue-sensitive. The main implication being, in turn, that those potential sources capable of placing the event in question into a larger context are regularly ignored, trivialized or marginalized (see also Wiegman et al. 1989; Nelkin 1995). Those sources able to address long-term environmental consequences, or speak to issues of mitigation and prevention, are shown to be routinely displaced from journalistic 'hierarchies of credibility' (Becker

1967). Where this does not occur, more often than not they are made to adapt their message to the discursive rules of inclusion and exclusion in operation (see also Anderson 1997; Chapman et al. 1997; Allan 1999; Allan et al. 2000; Smith 2000).

Nuclear nightmares

Several of the imperatives underlying the more salient 'news values' associated with 'extraordinary' environmental risks at the time were made all too apparent following the catastrophic events at Chernobyl in Ukraine. 'The Chernobyl era', as it has since been described by Russian commentators, began in the early morning of 26 April 1986 when a nuclear reactor overheated and exploded; 2 people were killed instantly (with 29 more to follow in the next year), massive fires broke out across the reactor complex and fissionable isotopes were released into the environment. The particular weather conditions helped to send a radioactive plume of gas and particles over 1500 metres into the atmosphere (Bunyard 1988: 35), promptly diffusing its radiation across the whole of Europe and beyond in the form of nuclear rain. For European countries, the first sign that something was wrong occurred two days later when workers at a nuclear reactor in northern Sweden registered abnormally high levels of radiation emanating from a source they could not immediately determine.

Official news of the accident was very slow to emerge from Tass, the Soviet news agency in Moscow. It was nearly 72 hours after the disaster before the first reference to it appeared, and even then it contained little by way of information (local radio announcements, in contrast, had been made at the start of the evacuation of Pripyat after about 36 hours). The afternoon edition of *Izvestia* on 29 April reproduced the official announcement:

> An accident has taken place at the Chernobyl power station, and one of the reactors was damaged. Measures are being taken to eliminate the consequences of the accident. Those affected by it are being given assistance. A government commission has been set up.
>
> (cited in McNair 1988: 140)

Many Soviet officials were quick to discredit Western news agencies for exaggerating the scale of the problem, insisting that they were deliberately distorting the facts in the interest of promoting anti-Soviet propaganda. It is indeed fair to say that much of the Western news coverage was anything but 'journalistically balanced' in its appraisal of the situation. In the US, the three main televisual networks and several major newspapers reported that over 2000 people had died. The *New York Post* went even further, publishing the headline: 'MASS GRAVE – 15,000 reported buried in Nuke Disposal Site'. Dominant themes identified in the coverage included representations of Soviet officials as having scant regard for human life. Further, they were presented as being committed to concealing the truth about the accident, and being hopelessly reliant on backward, primitive nuclear technology (Friedman et al. 1987; Luke 1987; Rubin 1987; Wilkins and Patterson 1987; McNair 1988).

In retrospect, a more significant issue emerges than the one signalled by journalists engaging in Cold War politics. Specifically, to the extent that the resultant news coverage framed the crisis at Chernobyl as a 'freak accident', a 'one in a million technical blunder', it was contributing to the implicit *normalization* of nuclear power as a safe source of energy. These types of news accounts, as Luke (1987: 153) argues, served to 'stress the strengths of the status quo, glossing over the accident that has torn only a small, temporary hole in the conventional order'. This framing of the crisis was made easier to sustain, in part, because previous nuclear accidents had transpired under conditions of secrecy. Included here are crises at Chalk River, Canada (1952), Windscale in the UK (1957), Kyshtym in the Ural mountains (1957), as well as in Idaho (1961) and just outside Detroit (1966) in the US (see Luke 1987; Spinardi 1997). For many reporters covering the Chernobyl crisis, then, the only point of reference was the partial meltdown which took place at the Three Mile Island nuclear power plant in the US in March 1979 (see Sandman and Paden 1979; Friedman 1981; Mazur 1981; Stephens and Edison 1982; Cunningham 1986; Rubin 1987). The similarities between how the two nuclear accidents were covered are striking, leading Rubin (1987: 42–3) to point out, for example, that officials in charge 'were quick to put the best face on developments and reluctant to confirm bad news. This demeanor diminished their credibility and reduced the number of potentially trustworthy sources with firsthand knowledge of the accident to few, or even none.' Moreover, he adds, 'certain kinds of information – particularly the amount of radiation released beyond the confines of the plant – were simply never provided or were provided too late to assist a worried public' (Rubin 1987: 43). Indeed, as Giddens (1998: 28–9) observed following the tenth anniversary of the Chernobyl disaster: 'No one knows whether it is hundreds – or millions – of people who have been affected by the Chernobyl fall-out.'

Many of the deficiencies indicative of Western news coverage of post-Chernobyl developments in nuclear energy are attributable to the journalistic search for the novel and the unusual, for dramatically compelling 'news pegs' confinable within episodic narratives. This is not to suggest that British or North American journalists, for example, are failing to report on various problems with these associated with nuclear power technologies. Typical of this coverage, however, is an emphasis on specific events, such as accidental 'leaks' or 'spills', to the detriment of a thorough accounting of the embodied risks for citizens over a period of time longer than yesterday's headline. The parameters of public debate over nuclear power are now largely defined by disputes among 'experts' regarding how best to 'administer' and 'regulate' these technologies. Despite the concerted efforts of anti-nuclear voices, not least those associated with groups like Greenpeace, the news audience is regularly asked by the technology's advocates to accept that nuclear power is 'clean', 'efficient' and 'harmless'. References to 'environmental friendliness' are typically coupled with those of 'cost effectiveness', the aim being to displace counter-arguments for alternative, non-nuclear technologies as being both 'impractical' and 'uneconomical'. In this way, the various rationales mobilized with the intent of legitimizing a continued reliance on nuclear technologies appear to be evermore firmly entrenched in a 'common sense' of **nuclearism**, one endorsed to varying degrees by the news media (see also Irwin et al. 2000; Welsh 2000).

To further elucidate the contours of this 'common sense', it is worth pausing to briefly consider a contrary instance, where its organizing tenets are disrupted. More precisely, in examining an *extraordinary* event, where the typical types of claims made about nuclear power as being 'safe', 'clean' and 'cheap' are challenged, the *ordinariness* of this 'common sense' is better discerned. Here attention turns to the British tabloid press's coverage of a so-called 'criticality accident' at the uranium processing plant in Tokaimura, Japan on 30 September 1999. To date, this has been the worst nuclear accident since Chernobyl. It led to the death of 2 plant workers and more than 40 others being treated for exposure to high levels of radiation. By reading this coverage 'against the grain', so to speak, it is possible to pinpoint several discursive strategies which have been utilized, knowingly or not, to help normalize the uncertainties of nuclear risk. First, though, the headlines:

> THE NUCLEAR NIGHTMARE – 300,000 forced to flee after disaster at Japanese plant
>
> (*Express*, 1 October 1999, p. 1)
>
> NUCLEAR LEAK IS OUT OF CONTROL
>
> (*Daily Mail*, 1 October 1999, p. 1)
>
> Nuclear leak 'is out of control'
>
> (*The Sun*, 1 October 1999, p. 1)

Nuke leak hurts 19

(*Daily Star*, 1 October 1999, p. 2)

Brits flee nuke leak

(*Mirror*, 1 October 1999, p. 9)

Even at the level of the headlines, it is possible to discern important differences in the respective newspapers' mode of address. Varying degrees of formality emerge around the term 'nuclear', for example, with two of the papers judging the everyday vernacular 'nuke' to be more appropriate for their readers. The word 'nuke' may signal for some a more comfortingly populist stance, as the word registers ideologically outside of the officially sanctioned, expert-endorsed lexicon. Similarly noteworthy in this regard is the word 'leak' to characterize the accident in four of the headlines, arguably calling forth associations which fail to attend to the specificity of the crisis (note the awkward juxtaposition of 'leak' with 'hurt'). Also, the headlines divide into distinct categories – three prioritize the implications of the 'leak' for people, while the remaining two pinpoint the issue of control (or absence of it) as the most newsworthy feature.

Even more telling, however, are the ways in which different news sources are handled in the different newspapers' news accounts. Looking at each account in turn, for example, it is possible to show how the respective journalists have imposed a normative order on those sources deemed sufficiently 'credible' or 'legitimate' to be quoted or paraphrased. That is to say, the standard inverted pyramid style format of hard news dictates that the most newsworthy sources are placed in the account's opening paragraphs, and those judged least newsworthy in the latter ones (if at all) in descending order of significance. This 'hierarchy of credibility' (Becker 1967), in turn, is likely to influence the readers' interpretation of the account. Those voices heard at the outset are far better able to set down the primary definition or terms of reference for the event in question, with those voices which follow having to be inserted within these limits accordingly (see also Hall et al. 1978; Allan 1999). To take one example from the stories cited above, the *Express*'s front-page item 'The Nuclear Nightmare' draws upon the following sources, either quoting, paraphrasing or attributing a speaking action to them, in this order: 'the Japanese government', 'Chief Cabinet Secretary Hiromu Nonaka', 'ministers', 'spokesman for British Nuclear Fuels', 'state official', 'doctors', 'the government', 'government spokesman Hiromu Nonaka', 'Japanese government', 'President Bill Clinton', 'anti-nuclear activist Pete Roche of Greenpeace'. All in all, the news account is 22 paragraphs long, spanning the first and second pages of the newspaper. Where these sources are called upon to characterize the use of nuclear power, each statement – to varying degrees – is pro-nuclear, with one exception. The exception is 'anti-nuclear activist Pete Roche of Greenpeace', whose statement is relegated to paragraph 22, the very last one of the account.

This line of analysis can be pursued much further, of course. The normative limits or parameters shaping how the reality of the crisis is to be defined can be thrown into even sharper relief, for example, by reordering the alignment of sources. That is to say, had the voice of Greenpeace appeared at the top of the hierarchy of source credibility, one would anticipate that a very different narrative would have ensued. Such a structure would likely have been regarded as being indicative of advocacy journalism, however, given that Greenpeace is likely to be seen as being 'political' in a way that voices from, say, the nuclear industry are not. Indeed, the 'spokesman for British Nuclear Fuels', fourth in the ordering of sources above, is called upon for a statement of apparent fact: 'Criticality is incredibly rare. There has only been one incident like this before, in America 50 years ago.' No effort is made in the account to challenge this assertion, or to question what definition of 'criticality' is being employed. It is similarly significant that blame for the accident is directly attributed to the workers involved ('The blunder happened as workers mixed uranium with nitric acid to make fuel') by the reporter. In contrast, the role of managers in properly supervising the maintenance of adequate safety standards escapes criticism. Moreover, no attempt is made to raise the issue of why the private company operating the facility, in a residential neighbourhood, was using a system in which the creation of a 'critical mass' of material was even possible. Instead, it is the lone voice of Greenpeace, in the final paragraph, which calls into question the safety of 'Japan's entire nuclear programme'. By the account's conclusion, then, and presumably despite the efforts of Greenpeace, no space has been given to question the risks inherent to nuclear power in general, or to acknowledge the possibility that what happened in Japan may very well happen elsewhere.

Public debates about nuclear power have recently taken on a new sense of urgency. 'September 11 has been the biggest challenge to nuclear power since Chernobyl,' argues Lochbaum, a nuclear engineer at the Union of Concerned Scientists in the US. 'Congressmen who have had very little interest in nuclear power in the five years I've been at U.C.S. are suddenly competing with each other to examine security issues at the plants' (cited in *The New York Times*, 5 December 2001). Heightened fears of terrorist attack underscore the risk that a nuclear power plant would not be able to withstand the impact of an airliner crash. In the aftermath of such an event, a 'Chernobyl situation' – to use the phrase of one US official – would likely arise whereby a nuclear meltdown would occur. The ensuing release of radiation would be catastrophic, with some industry scientists conceding that it would be far greater than the Chernobyl incident as it would likely prove lethal to anyone living within kilometres of the plant. In the words of the president of the Nuclear Control Institute in Washington: 'It is prudent to assume, especially after the horrific, highly coordinated attacks of September 11, that [the terrorists] have done their homework and are fully capable to attack nuclear plants for maximum effect.' He then added his view that '[t]here is a security vacuum out there now, a very dangerous vacuum' (cited in *The Guardian*, 26 September 2001). The relative vulnerability of each of the reportedly 103 nuclear power stations across the US is now being reappraised, with new risk assessments being made. In Britain, the nuclear reprocessing plant at Sellafield in Cumbria has been the subject of particular concern. 'What are very big risks are the huge tanks of very, very radioactive liquid stored in reprocessing plants,' one nuclear physicist told *The Guardian*. 'They contain a huge amount of radioactivity and are less-well protected than reactors, which are within very large concrete shields' (*The Guardian*, 18 September 2001). Suddenly, from one country to the next, the 'common sense' of nuclearism is threatening to start coming unravelled as a new politics of risk emerges.

In the next section, attention returns to environmental reporting more generally. The discussion draws, in the first instance, upon several pertinent insights into the day-to-day routines of newswork provided by science and environmental journalists.

Covering environmental issues

'No pictures, no story' is the way broadcast journalist Flatow (1997: 40) sums up the significance of the visual imperative for television news coverage of environmental issues. Without pictures, and preferably moving ones at that (video or film), she argues, it is difficult to convince a sceptical editor that the story should be covered. Television news reporter Ropeik (1997: 35) echoes this point with his observation that: 'Science stories, and environmental stories that often include science, are a tough sell to television news managers more used to more "hard" or "breaking" or, to be honest, more "sensational" stories' (see also Cottle 2000). Like Flatow, Ropeik believes that the power of pictures to tell a story is what can make television news coverage of science so effective. By way of example, he writes:

> We did an investigative series on serious carbon monoxide and nitrogen dioxide pollution at indoor skating rinks, caused by exhaust from ice resurfacing machines running without pollution control equipment in poorly ventilated buildings. No matter how I wrote that piece, the pictures were what really told the story: fumes from the resurfacing

> machines operating right in front of skating kids, and pictures of air-testing equipment registering 'evacuation' levels as we sat taking the tests among spectators in the stands.
>
> (Ropeik 1997: 35)

Needless to say, of course, sound environmental reporting needs more than telling pictures to make the story come alive for the audience. Foremost among the additional challenges facing the television reporter, in Ropeik's view, is the brevity imposed by the news story format – in his experience, about one minute and 40 seconds long is typical. This time restriction not only dramatically reduces the amount of detail that can be included in the news item, but also constrains the reporter's ability to place these details in an appropriate context. Ropeik (1997: 36) expresses his frustration in having to make compromises over the depth of his reporting, acknowledging that 'facts get left out. Big ones, sometimes.'

Even when the facts are left in, so to speak, difficulties arise with regard to the reporting of environmental statistics. Cohn (1997), a former science writer on a daily newspaper, offers a telling example from a debate over the Clean Air Act in the US:

> one issue was whether or not to require automakers to remove 98 percent of tailpipe emissions, compared with a previous 96 percent. The auto-industry called it an expensive and meaningless 2-percent improvement. Environmentalists said it would result in a 50-percent decrease in automobile pollution. The lesson: In any such reporting, seek all the figures, not just the assertion someone tries to spoonfeed us.
>
> (Cohn 1997: 108)

A problem with this kind of reporting, which Cohn suggests is especially pronounced where risk issues are concerned, occurs when 'one side tells us the sky is falling, and the other says it's not. Or here's the cure, no it's not' (Cohn 1997: 102). As he proceeds to elaborate:

> A politician says, 'I don't believe in statistics,' then maintains that 'most' people think such and such. Based on what? A poll says, 'Here's what people think,' with a 'three-point plus or minus margin of error.' Believable? A doctor reports a 'promising' new treatment. Is the claim justified or based on a biased or unrepresentative sample? An environmentalist says a nuclear power plant or waste dump causes cancer. An industrialist indignantly denies it. Who's right?
>
> (Cohn 1997: 102)

Issues about who to believe, what to report, and which points are worth pursuing, are anything but straightforward. Critically examining claims made by contending news sources, Cohn (1997: 103) argues, 'requires not so much an understanding of formulas as an understanding of the nature of rational evidence, the array of facts or probable facts that make us decide something is believable'. Not surprisingly, then, he makes the case for reporters adapting the methods of the scientist to judge claims of fact using similar rules of evidence. 'Separating the truth from the trash', to use his turn of phrase, means looking beyond the conflicting statements made by 'duelling experts' (Detjen 1997) so as to recognize the ways in which source credibility is managed. 'For all of us,' Cohn (1997: 109) maintains, 'telling the probable truth from the probable trash is to a large extent a matter of attitude, of healthy skepticism – not cynicism or believing nothing, but an informed kind of "show me!".'

Speaking as a science correspondent in radio journalism, Harris (1997: 166) argues that 'the "he-said-she-said" style of coverage is the easy way out, but it helps nobody'. In discussing his experiences covering environmental issues, he maintains that 'taking the side of the little guy has obvious appeal'. Indeed, he acknowledges that the first responsibility of a journalist is to them, but then proceeds to ask: 'what if they're misinformed? Where's our responsibility to the truth? What is the truth?' (1997: 166). Answers to questions such as these ones have over the years proven to be particularly vexing where environmental issues are concerned. 'In the early days,' Harris (1997: 167) recalls, 'it was easy to join the crusade and side with the Davids – the cash-starved environmental groups and outraged citizens – against the Goliaths, so often industrial giants that were using everyone's air and water as their own corporate sewers.' Today, though, things are not quite so simple, in his view. Environmental groups, he contends, have become 'multimillion-dollar operations, searching for the next big issue to draw in more membership dollars' while, at the same time, citizen groups 'may have more emotions than facts to back them up' (1997: 166). Meanwhile, the Goliaths, while still motivated by their 'own interests at heart', continue to argue – with some justification in Harris's opinion – that 'environmental regulations are costing consumers big bucks' (1997: 166; see also Bakir 2001).

The difficulties associated with negotiating conflicting claims from different sources can be compounded when the scientific basis of the respective claims being made is in contention. Trying to decide whether or not to report on a particular scientific development is often difficult, as Salisbury (1997) points out:

> When a given piece of research would make a good story, to what extent should a science writer act as a gatekeeper on scientific grounds? I'm not comfortable with the proposition that we should refrain from

> publicizing work that is controversial or politically incorrect. When we do so, however, it is important that we tell our news media colleagues and other readers that the work has strong detractors. On the other hand, I have no desire to promote flawed science. Unfortunately, there will always be cases where it is extremely difficult for [reporters] to tell the difference.
>
> (Salisbury 1997: 224)

The journalist on the environmental beat has the formidable task of translating highly technical information into jargon-free prose that is understandable to the audience. 'It's not uncommon', according to Detjen (1997: 179), 'for an environmental reporter to be writing about science, public health, business, the courts, government, even religion.' Too much of environmental journalism is shallow and incomplete, in his view, not least because the underlying political, legal and scientific processes are not addressed in adequate detail. Even when they are identified, it is hard work to discern how they are shaping the scientific claims being articulated by different sources. Indeed, as he advises other journalists:

> If you are writing about acid rain, don't just write that the rain in some parts of the Adirondack Mountains has a pH of 3. Translate that. Write that the rain has become as acidic as vinegar. Don't just write that *X* parts per million of sulphur dioxide are being released into the air. Compare those levels with those that can trigger asthmatic attacks in the sick and the elderly.
>
> (Detjen 1997: 177)

Consequently, a journalist engaged in reporting on environmental risk issues, according to Harris (1997), needs to approach a given story from three different angles simultaneously. 'First, it's a human story involving deep-seated emotions about fear and health and society; it's a story about power and economics; and finally, it's a story rooted in science – and imperfect science at that' (Harris 1997: 167). Such an approach, in his view, might make an environmental news story more difficult to write, given that the 'easiest stories to tell have heroes and villains, victims and evil or avaricious perpetrators', but it will necessarily attend to the complexities involved more effectively. At issue is the importance of resisting the temptation to squeeze an environmental story into a rubric strictly defined by the motivations (and funding) of the different sources engaged in the dispute.

Similarly tempting is the tendency to overstate the potential severity of the environmental risk under scrutiny. Newspaper journalist Rensberger (1997: 15) argues: 'The tradition in environmental writing, unfortunately, is to be rather alarmist in tone. Environmental stories are usually bad news, warnings of impending disaster.' At the same time, the tendency for the news media to represent an environmental crisis as a specific event-oriented catastrophe, as opposed to recognizing that it is an outcome of bureaucratic calculations and decisions, has been well documented by researchers. As Wilkins and Patterson (1990) maintain:

> While risk analysis indicates that not all risks are alike, news media coverage of a variety of hazards and disasters tends to follow predictable patterns. Neither the unpredictability nor the high degree of complexity of hazards fits neatly into a newsgathering process that places a high priority on meeting deadlines. Therefore, news about hazards often is moulded to the medium. A day-long debate about the location of a toxic dump is reduced to 30-second 'sound bites' from each side and footage about angry demonstrators staking 'pseudo-events' for the benefit of the cameras. In the end, the audience is *entertained* by the hazard without being *informed about it*.
>
> (Wilkins and Patterson 1990: 13, original emphasis)

These types of factors have been discernible in a variety of instances of news coverage concerned with other environmental disasters during the mid-1980s and since. Particularly catastrophic was the 1984 chemical leak at the Union Carbide pesticide plant in Bhopal, India which reportedly killed over 2500 people in a matter of days (Wilkins and Patterson 1987; Salomone et al. 1990). A further instance is the oil spill of the *Exxon Valdez* in the Alaskan waters of Prince William Sound in 1989, at the time the largest spill in US history. In their critique of the 'disaster narrative' which emerged in the reports of some newspaper journalists, Daley and O'Neill (1991: 53) maintain that it 'naturalised the spill, effectively withdrawing from discursive consideration both the marine transport system and the prospective pursuit of alternative energy sources'. As a result, they contend, attendant environmental issues were directed 'away from the political arena and into the politically inaccessible realm of technological inevitability', thereby reproducing 'the political and corporate hegemony of Big Oil' (Daley and O'Neill 1991: 53; see also Smith 1992; Hansen 2000; Bakir 2001).

Public concern about environmental issues in countries like Britain reached new levels in the latter part of the 1980s, although it had begun to dissipate by the early 1990s. This perception has been based on studies of news coverage (Hansen 1991, 1993; Gaber 2000), as well as by considering related types of evidence. Among the latter are opinion poll surveys, membership figures for environmental organizations and the advent of 'green' consumerism with the launch of new 'environmentally friendly' products in shops (Gauntlett 1996; Coupland and Coupland 2000; see also Darier 1999; Harré et al. 1999; Wykes 2000). Judging from the vantage point of today, however, popular interest in environmental issues appears to be once again on the increase, a dynamic evidently owing much to public alarm over BSE or 'mad cow disease', genetically modified crops, and the **foot-and-mouth-disease** outbreak in 2001 (see Chapter 7). In the US, there appears to be something of a consensus that environmental concerns have continued to grow steadily since the 1960s. 'In the 1960s,' recalls Detjen (1997: 173), an environmental journalist in the US, 'these issues were small eddies on the edge of a river flowing through our lives and our society; today they have become one of the river's most powerful currents.' Not surprisingly, the public's interest in environmental news has tended to wax and wane over these years. Detjen (1997) maintains that the 'prevailing mood in the country' is a decisive factor in this regard. 'During periods of economic uncertainty people focus more on jobs and the adequacy of health insurance and other financial issues,' he argues, while subsequent to major environmental calamities 'public interest in environmental issues increases rapidly' (Detjen 1997: 173–4).

In recognizing how difficult it is to gauge public interest in media reporting of environmental issues, some commentators are looking for explanations in the apparent globalization of environmental risks, threats and hazards (see Guedes 2000; Szerszynski and Toogood 2000; van Loon 2000). Here references are frequently made to a growing public awareness of the intricate ways in which environmental concerns are seen to impinge upon everyday life in 'first world' countries. As Lacey and Longman (1997) maintain:

> As the debate has matured it has become apparent that the problems are more complex and involve 'us' as well as 'them'. The destruction of the tropical rainforests involves poverty, exploitation, international debt and the demand for tropical hardwoods from Western nations. The problem of global warming involves the vast consumption of fossil fuels by Western nations for heating, transport and manufacture. There are no easy solutions to these problems.
>
> (Lacey and Longman 1997: 115; see also Bell 1994; Wilson 2000)

At stake, then, is the need to question anew the cultural politics of environmental risks precisely where the global meets the local, where proclaimed divisions between 'us' and 'them' are shown to be not only misleading, but also dangerous. A sense of urgency informs this kind of research as around the globe the specific ways in which the media encourage their audiences to think about environmental risks are increasingly becoming the subject of public controversy. As Leiss and Chociolko (1994) maintain:

> Charges of media bias or sensationalism, of distorted or selective use of information by advocates, of hidden agendas or irrational standpoints, and of the inability or unwillingness of regulatory agencies to communicate vital information in a language the public can understand, are common. Such charges are traded frequently at public hearings, judicial proceedings, and conferences, expressing the general and pervasive sense of mistrust felt by many participants towards others.
>
> (Leiss and Chociolko 1994: 35–6)

At the same time, of course, there is some room for optimism. 'The mounting reliance of everyone in modern society on the judgements of "experts" ', as Grove-White (1998: 50–1) observes, 'is paralleled by the growing ability of many of us, reinforced by modern media, to deconstruct political reassurance couched as scientific or technical "fact".' This difficult work of deconstruction, as this chapter has sought to show, necessarily entails challenging taken-for-granted assumptions about the presumed neutrality of science as a dispassionate 'set of facts' where the calculation of risk is concerned.

To close, then, news coverage of environmental risks matters. 'Perhaps more than most stories,' as Stocking and Leonard (1990: 42) argue, 'it needs careful, longer-than-bite-sized reporting and analysis, now.' Public demands for the news media to improve their coverage do appear to be gaining ground in some places. As worries about declining audience figures are expressed in newsrooms, 'the environment' is perceived to be the sort of 'soft' news story which attracts the attention of those members of the public – especially young people – who often do not regularly follow the news (see also Gauntlett 1996; Cottle 2000; Wykes 2000). Also, it is arguably the case that journalists are becoming more self-reflexive about the ways in which they represent environmental risks, a development informed in part by pressures brought to bear in the name of 'civic' or 'public' journalism (see also Rosen 1999). Pressure to change is also coming from those environmental activists who recognize that while the media are playing a crucial role in sustaining the imperatives of 'expert' risk assessment, they are also creating spaces, albeit under severe constraints, for counter-definitions to emerge.

Reporting which reduces environmental risks to isolated events or incidents devoid of adequate context, where 'personalities' engaged in disputes are made to stand in for larger economic, political and cultural factors, fails to make the necessary connections at a social structural level. Reporting which *is* sensitive to the lived contingencies of risk, in sharp contrast, takes these necessary connections as its starting point.

Note

1 An earlier version of some portions of this chapter's discussion appeared as part of my contribution to the 'Introduction' to Allan, S., Adam, B. and Carter, C. (eds) (2000) *Environmental Risks and the Media*. London and New York: Routledge. I wish to thank my co-editors, as well as Routledge, for permission to draw upon some of this material here.

Further reading

Allan, S., Adam, B. and Carter, C. (eds) (2000) *Environmental Risks and the Media*. London and New York: Routledge.

Anderson, A. (1997) *Media, Culture and the Environment*. London: UCL Press.

Beck, U. (1995) *Ecological Politics in an Age of Risk*. Cambridge: Polity.

Chapman, G., Kumar, K. Fraser, C. and Gaber, I. (1997) *Environmentalism and the Mass Media: The North–South Divide*. London: Routledge.

Franklin, S., Lury, C. and Stacey, J. (2000) *Global Nature, Global Culture*. London: Sage.

Irwin, A. (2001) *Sociology and the Environment*. Cambridge: Polity.

Lash, S., Szerszynski, B. and Wynne, B. (eds) (1996) *Risk, Environment and Modernity*. London: Sage.

Macnaghten, P. and Urry, J. (1998) *Contested Natures*. London: Sage.

Neuzil, M. and Kovarik, W. (1996) *Mass Media and Environmental Conflict: America's Green Crusades*. Thousand Oaks, CA: Sage.

Smith, J. (2000) *The Daily Globe: Environmental Change, the Public and the Media*. London: Earthscan.

6 BODIES AT RISK: NEWS COVERAGE OF AIDS

> Journalists take great pride in their ability to detach from the news. The media maintain an invisible boundary to keep the news at bay so they can maintain their objectivity. AIDS recognized no such boundaries. As journalists began to contract it, AIDS became increasingly difficult for the media to ignore.
>
> (Edward Alwood, journalist)

Readers of *The New York Times* newspaper on 3 July 1981 may have been perplexed by a short item on page 20 of that day's edition concerning a strange new syndrome. The news item in question, headlined 'Rare Cancer Seen in 41 Homosexuals', described the results of a medical study into several cases of Kaposi's Sarcoma occurring in otherwise healthy, young gay men. The medical reporter, Lawrence K. Altman (also a physician), regarded the outbreak to be puzzling enough to warrant attention, but understandably appeared to have had no inkling of just how newsworthy it would prove to be. 'The cause of the outbreak is unknown', he wrote, 'and there is as yet no evidence of contagion.' The focus of the item then turned to the possible significance of the study's scientific findings for determining the causes of more common types of cancer. Following this first news story, only two further mentions of the spread of the disease would appear on the pages of the newspaper by the end of 1981 (see also Kinsella 1989; Alwood 1996).

Related developments, although not recognized as such at the time, had been receiving limited news coverage over the previous month. Specifically, a handful of news items had appeared in the US press following a Centers for Disease Control (CDC) report, titled 'Pneumocystis Pneumonia – Los Angeles', published on 5 June 1981. The CDC report aimed to alert the US medical community to a deadly new disease, which had crippled the immune systems of five previously healthy men. These highly unusual cases of Pneumocystis carinii pneumonia were thought to be linked, to use a phrase from an Associated Press (AP) wire service item, to 'some aspect of a homosexual lifestyle'. Eventually, in the months to follow, this new type of pneumonia would be considered in relation to the Kaposi's Sarcoma cases. Some scientists had begun to wonder if there might be a common underlying factor at work. No one knew what that factor might be, however, and the puzzle failed to sustain interest among science and medical journalists.

Substantive indications of the impending epidemic would begin to consolidate in medical journals only over the course of 1982. References to a possible 'gay cancer' or 'gay-related immune deficiency' (GRID) were appearing on a regular basis. Towards the end of that year, as the potential seriousness of this public health problem slowly became more pronounced, journal articles began to use the term Acquired Immune Deficiency Syndrome or AIDS to describe the mystifying condition. By the year's end, the Centers for Disease Control in the US reported 800 cases and 350 deaths from AIDS. Coverage in the mainstream news media was virtually non-existent. Efforts made by some journalists working in the gay press to push the issue onto the national news agenda were met with indifference. This silence – deafening to those diagnosed with the disease and their loved ones – was abruptly broken in May 1983. An editorial published that month in the *Journal of the American Medical Association* (*JAMA*), by Dr Anthony Fauci, director of the National Institute of Allergy and Infectious Diseases, raised the startling proposition that AIDS might prove to be transmissible to the entire population via 'routine close contact'. Suddenly, it appeared, everyone was at risk of this deadly new affliction. 'Defined as a homosexual disease,' Nelkin (1995: 102) writes, 'AIDS had attracted little public attention: when it seemed that it might extend beyond the gay community, coverage significantly expanded.'

The precise origin of AIDS, as well as the means of its transmission, quickly became the subject of intense speculation. In the absence of satisfactory scientific explanations, as Grimshaw (1997) notes, a range of 'lay' or 'folk' beliefs about origins and causes vied with one another for public attention. He identifies four such beliefs as being particularly prominent:

- *endogenous*: the belief that an individual has some innate characteristic or flaw that causes them to have disease (the identification of all the initial cases amongst gay men led to suggestions that homosexuality 'caused' AIDS).
- *exogenous*: the belief that disease lurks opportunistically in the external environment, 'waiting to pounce'; this is linked to ancient beliefs that diseases spread like a miasma through the air, beyond human control.
- *personal responsibility*: the belief that a person's behaviour makes it acceptable or appropriate that they should become diseased. In

the context of HIV, this underlies distinctions between 'innocent' victims, such as children, and 'guilty' victims, such as gay men, injecting drug-users, prostitutes and the sexually promiscuous.

- *retributionist*: the belief that actions which infringe supposedly fundamental moral values are likely to bring about divine retribution. AIDS was considered by some to be divine retribution for transgression against certain biblical proscriptions on homosexuality and promiscuity.

(Grimshaw 1997: 379–80)

The ideological purchase of these beliefs was such, Grimshaw argues, that it was almost inevitable that 'social responses to people with AIDS would be characterized by prejudice (literally "forming judgments without proper information") and discrimination' (Grimshaw 1997: 380). By the time of the scientific breakthrough leading to the isolation of HIV as the causative agent of AIDS – the precise credit for which is still a matter of dispute between rival French and US laboratories – much of the symbolic damage had been done. Due in part to the ways in which the disease was portrayed in the media during this early phase, Grimshaw (1997: 381) maintains, to this day it 'remains perhaps irrevocably stigmatized'.

This process of stigmatization is complex, and has unfolded unevenly in a multiplicity of ways in diverse situational contexts. Definitional contests over the realities of AIDS have given voice to deep-rooted fears, suspicions and anxieties, many of which were expressed in the pernicious language of homophobia in some contexts, racism in others. In trying to map the broader contours of this process, researchers frequently draw upon a concept of 'moral panic' (Cohen 1972) to try to elucidate its features. At root, as McRobbie (1994) observes, a moral panic is about 'instilling fear in people'. In so doing, she argues, people are encouraged 'to try to turn away from the complexity and the visible social problems of everyday life and either to retreat into a "fortress mentality" – a feeling of hopelessness, political powerlessness and paralysis – or to adopt a gung-ho "something must be done about it" attitude' (McRobbie 1994: 199; Critcher, in press). In the next section, our discussion turns to the emergent moral panic around HIV/AIDS issues, paying particular attention to the media as sites of public deliberation and debate about the attendant risks for their readers, listeners and viewers.

An im/moral panic

Today, HIV (human immunodeficiency **virus**), the virus believed to cause AIDS, is one of the worst epidemics in the history of humankind. Official estimates vary, but it is generally understood that around the world over 21 million people have died from the condition, and over 36 million more are now infected (Altman 2001a). About 15,000 new cases are being identified daily (Watanabe 2001) in a prevailing climate of public fear, persecution and recrimination – their chances of obtaining adequate care and treatment thereby being diminished as a result. 'What is so very striking about the moral panic around AIDS', writes Weeks (1985: 45), 'is that its victims are often blamed for their illness.'

Critical research on media portrayals of HIV/AIDS suggests that much of the news coverage has contributed to the creation of this hostile climate – typically characterized as a 'moral panic' – around the disease. While some journalists refused to succumb to sensationalist reporting, far too many others produced news accounts which effectively dehumanized people living with the condition. The emergence of this public health crisis was formulated from the outset by some media commentators in bitterly divisive terms, most notably by the insistence on describing it as the 'gay plague'. In the absence of informed, incisive reporting, narratives of moral outrage were mobilized across the media, some inciting fear, even hatred. Watney (1997: 43) argues that gay men were made to 'stand outside the "general public", inevitably appearing as threats to its internal cohesion'. This cohesion, he adds, 'is not "natural," but the result of the media industry's modes of address – targeting an imaginary national family unit which is both white and heterosexual' (Watney 1997: 43). Moral panics provide the 'raw materials', that is, the words and images, by which different moral constituencies encourage people to identify with their preferred definitions of the crisis. This process of identification is, as noted above, deeply politicized. Newspaper and television reports consistently refuse the possibility of identification *with* gay men: 'The gay man is effectively and efficiently positioned as he-with-whom-identification-is-forbidden' (1997: 12).

Accordingly, Watney maintains that in examining the early stages of the moral panic around AIDS, it is crucial to place sufficient emphasis on the 'overall ideological policing of sexuality', especially where issues of representation are concerned. In deconstructing the forms of this policing, the 'ideological confrontations' which have led to the scapegoating of gay men need to be discerned. The media's framing of AIDS as a 'gay plague', he argues, implicitly invited members of the public to believe that the syndrome was 'a direct function of a particular sexual act – sodomy – and, by extension, of homosexual desire in all its forms. Thus, by contingency, even lesbians are suspect and newly newsworthy' (1997: 12). Central to the discourse of AIDS promulgated by the media, according to Watney (1997: 12), was what he terms the rhetorical figure of 'promiscuity', that is, 'as if all non-gays were either monogamous or celibate and, more culpably still, as if Aids were related to sex in a quantitative rather than qualitative way'. In this manner, then, 'promiscuity' became routinely employed by the media as 'a sign of homosexuality itself, of forbidden pleasures, of *threat*' (1997: 12). At stake, in order to uphold this AIDS discourse, was the need to 'contain the dangers of deviance, henceforth to be branded indelibly with the ideological skull-and-crossbones sign of Aids' (1997: 12–13). Drawing a parallel with how 'menstruation was used for centuries as "evidence" of Fallen Eve in every woman', Watney suggests that AIDS was being 'recruited as the natural, just and due reward for sex outside marriage' (1997: 13).

The extent to which various media discourses contributed to the establishment – as well as the policing – of certain ideological limits or boundaries around AIDS is similarly explored by Weeks (1993). The concept of 'moral panic', when used with sufficient care, is in his view 'a helpful heuristic device to explore the deeper currents which shaped the developing HIV/AIDS crisis' (Weeks 1993: 25). Of the multitude of moral panics which have emerged over previous decades, he argues, those which have intertwined moral and sexual issues have been particularly acute. In the case of AIDS-related illnesses, the intense fears and anxieties associated with the attendant uncertainties were at least partly understandable, yet the denigration of afflicted individuals and groups had tragic consequences. As Weeks points out, for example, there was appalling evidence of certain 'practices of decontamination' (a phrase borrowed from Sontag 1989) against both gay men and lesbians:

> restaurants refused to serve gay customers, gay waiters were sacked, dentists refused to examine the teeth of homosexuals, technicians refused to test blood of people suspected of having AIDS, paramedics fumigated their ambulances, hospitals adopted barrier nursing, rubbish collectors wore masks while collecting garbage, prison officers refused to move prisoners, backstage staff in theatres refused to work with gay actors, distinguished pathologists refused to examine bodies, and undertakers refused to bury them.
>
> (Weeks 1993: 26)

Each of these incidents, Weeks maintains, can be documented by looking at the press coverage of the time, some of which bordered on the hysterical. While acknowledging that they were not universal experiences – 'there was altruism, self-sacrifice and empathy as well' – he makes the point that 'all these things happened, to people vulnerable to a devastating and life-threatening disease; and the vast majority of these people were homosexual' (Weeks 1993: 26).

As the divergent strands of the moral panic around the 'gay plague' began to coalesce during the early 1980s, it was slowly expanding to encompass a range of other groups. While gay men made up the most of the reported cases, other groups identified as being at particular risk included intravenous drug users, haemophiliacs, and Haitian immigrants into the US. In a study of the experiences of the latter, Dubois (1996) argues that 34 of the first 700 patients to be diagnosed with AIDS in the US were Haitian-American. Due to the apparent lack of 'high-risk' factors associated with the other three groups, the CDC elected to define them as a distinct category. Justification for this categorization, according to Dubois, was based on the logic that those Haitian-Americans struck with the disease did not appear to belong to the other high-risk categories. That is, the assumption was made that 'transmission of the virus must have occurred through heterosexual contact; therefore, it was theorized, Haitians in general were more at risk than heterosexuals of other ethnic groups' (Dubois 1996: 24–5). This assumption was based on several misunderstandings, not least the cultural and linguistic contexts within which the interviews with the patients were conducted. 'It is doubtful', Dubois (1996: 25) contends, 'that the patients would have admitted to homosexual practices or drug use in front of a white doctor in a government hospital, especially given the generalized mistreatment of Haitian-Americans at the hands of American officials [and] the fear of deportation'. Later research indicated that among the majority of the patients, typical risk factors had been involved.

Before this type of evidence was officially recognized, however, the people of Haiti found themselves caught up in the pernicious forms of stigmatization associated with the larger moral panic. Media discourses about the ongoing search for the source of the condition routinely drew upon a language of blame and censure which, Dubois suggests, effectively contributed to the dehumanization of Haitians. Scientific and journalistic articles affixing blame on them for AIDS, he argues, typically invoked ethnocentric attitudes which denigrated their values, customs and traditions. According to some, the island's people had first become 'infected with the virus through ingestion of uncooked animal blood during a ceremonial sacrifice' (Dubois 1996: 25). Haitians were recurrently being characterized as 'backward', 'primitive', 'superstitious' and 'disease-ridden'. In this context, it is not surprising that some Haitian scientists, government officials and others reacted by insisting that the US itself was the source of AIDS. Allegations were made, for example, that tourists from that country had introduced the condition to the island – gay tourists, in particular, were singled out for blame – rather than the other way around. What quickly became certain, in any case, was that Haitian tourism, evidently the second largest source of foreign earnings before 1982, promptly began to collapse. 'Instead of bringing healing to Haiti,' Dubois (1996: 29) writes, 'the search for an origin sent another curse to the island.' The backlash of the AIDS scare had made an impoverished country's plight even more desperate.

Divergent theories regarding the origin of the AIDS virus, each advanced by its own constituency, competed vigorously for media attention around the world in the early 1980s. Just as Haiti was widely deemed by some to be responsible for the outbreak, others pointed the finger of blame at countries in Africa (a disease originating in monkeys? African swine fever?). Still others condemned scientists for manufacturing the virus. For example, a front-page story in the British *Sunday Express*, and reprinted around the world, declared: 'The killer AIDS virus was artificially created by American scientists during laboratory experiments which went disastrously wrong – and a massive cover-up has kept the secret from the world until today' (cited in Sontag 1989: 53). Related charges held the US Defense Department culpable – for others, it was the Central Intelligence Agency (CIA) – maintaining that the virus had been developed in 'germ warfare' research with the deliberate aim of targeting gays and black people. Each of these theories, and many more like them, found their way into public circulation via media organizations scrambling to make sense of the emerging epidemic.

Reporting the plague

Media discourses, like those of science, medicine and the others on which they routinely draw, construct certain preferred ways of talking about the lived realities of HIV/AIDS. In her book *AIDS and its Metaphors*, Sontag (1989) points out that from its earliest uses the term AIDS was made to represent an illness with a single cause, when in actuality it is a medical condition. As she writes, in 'contrast to syphilis and cancer, which provide prototypes for most of the images and metaphors attached to AIDS, the very definition of AIDS requires the presence of other illnesses, so-called opportunistic infections and malignancies' (Sontag 1989: 16). Describing the 'metaphoric genealogy' of AIDS, she points out that as a micro-process it tends to be described as an invasion (as cancer is), while when the focus is on its transmission, metaphors of pollution (reminiscent of syphilis) are typical. Military metaphors similarly abound, many with a 'science fiction flavour'. The virus is said to 'penetrate' **cells** so as to 'trigger' the replication of an 'alien product', leading to an 'all-out attack' on the immune system.

Crucially, these types of metaphoric flourishes, Sontag argues, are inextricably linked to an imputation of guilt:

> Indeed, to get AIDS is precisely to be revealed, in the majority of cases so far, as a member of a certain 'risk group', a community of pariahs. The illness flushed out an identity that might have remained hidden from neighbors, job-mates, family, friends. It also confirms an identity and, among the risk group in the United States most severely affected in the beginning, homosexual men, has been a creator of community as well as an experience that isolates the ill and exposes them to harassment and persecution.
>
> (Sontag 1989: 25)

AIDS, in some discourses, is held to be the product of a particular 'lifestyle', one characterized by 'unsafe behaviour' among those who are 'weak', 'indulgent' and 'delinquent'. By this logic, alleged addictions to 'deviant', 'perverse' sex, or to illegal chemicals (or both), mean that AIDS is a calamity one brings on oneself. Sontag, in briefly tracing the history of other diseases such as syphilis, shows how AIDS has revived similar phobias and fears of contamination among the so-called 'general population' across society. AIDS is deemed to be retributive, a punishment for an individual's illicit transgression, while also threatening to be collectively invasive for the 'morally upright'. Hence, in part, the reason why 'plague' quickly became the principal metaphor by which the AIDS epidemic was to be understood, in her view. The word 'plague', derived from the Latin *plaga* (stroke, wound), used in conjunction with AIDS revives 'the archaic idea of a tainted community that illness has judged' (1989: 46). In sustaining this sense that AIDS constitutes a moral judgement on society, she contends, the plague metaphor serves the interests of those seeking to promote and exploit public fear: AIDS becomes 'God's punishment', 'the revenge of nature' or 'a terrorist's weapon'.

Perhaps the most insidious dimension of the plague metaphor was the way in which it contributed to the normalization of an ideological dichotomy between alien, less than human 'others', on the one hand, and everyone else, on the other, during the initial stages of the moral panic. In the case of news reporting, the underlying imperatives of this 'us' versus 'them' dichotomy began to cohere as a form of prejudice from the outset. Mainstream journalists and their editors, in sharp contrast with their counterparts in the gay press (especially publications such as the *New York Native* and the *Advocate*), refused to see how a story ostensibly revolving around homosexuals and injecting drug users could possibly interest their audiences. The growing epidemic failed the test of newsworthiness so easily passed in previous years by certain other illnesses, such as Legionnaires' disease and toxic shock syndrome, which had received extensive coverage. Bishop, a medical reporter for the *Wall Street Journal* at the time, remembers why this was the case:

> The reason that the toxic shock and Legionnaires' diseases got so much coverage was that the editors felt both of those things threatened them and their families, and here AIDS did not – they didn't feel that it did, anyway. When editors don't feel threatened – when something doesn't threaten them, or their families, or the people they know – then it's not seen as a big story. Nobody ever said anything about the homosexual angle, but it was clear that they didn't know how to handle the intimation that it was being passed by sexual practices among gay men.
>
> (cited in Alwood 1996: 217)

This kind of prejudice – much of it allowed to remain at a tacit, seemingly 'common sensical' level in newsroom discussions – meant that the condition remained virtually ignored by the mainstream media. Only a small number of news organizations sought to disrupt this emergent consensus that AIDS was a health problem that only affected certain 'deviant' minorities. Even then, however, pertinent news stories were routinely downplayed or 'spiked' altogether, sometimes over little more than fears that they might upset the audience's sense of moral decency ('we are a family newspaper', evidently being a typical refrain among some editors).

AIDS became front-page news in the US on 31 May 1982 when the *Los Angeles Times* newspaper published an item titled 'Mysterious Fever Now an Epidemic'. This was the first occasion, according to Alwood (1996), that news of the disease received front-page treatment. As he observes, despite the fact that at the time new cases were being identified by medical authorities at a rate of one per day, 'most editors at newspapers nationwide did not consider gay deaths news' (Alwood 1996: 219). Garrett (1997), a medical reporter, similarly believes that AIDS was initially ignored by the US's mainstream journalists because the numbers of people affected were judged to be too small, and their locations too scattered, to warrant much attention. At the same time, however, also informing these criteria of newsworthiness, she argues, were certain 'morally dubious' assumptions being made on the basis of who was being afflicted. According to Garrett:

> Homophobia, fear of IV drug users, and dismissive attitudes towards Haitians and Africans allowed some journalists to convince themselves that the story was still small, even as the US death toll grew to 50,000, because the epidemic hadn't reached 'the general public'. It is inconceivable that news organizations would have allowed such a 'general public' issue to cloud coverage of, say, 50,000 dead Americans who

> succumbed over a period of four years, all of whom were elderly people of Scandinavian descent.
>
> (Garrett 1997: 159)

Garrett further recalls how the news media succumbed to a panic and fear response from the early days of AIDS. 'Reporters were afraid to touch their gay sources; camera operators refused to go on television shoots involving AIDS; some journalists reported sympathetically on vigilante actions and protests leveled by communities against gay men and children with AIDS' (Garrett 1997: 158). Meanwhile other reporters, she adds, 'wore gloves during their interviews and washed up as soon as they left the rooms of AIDS patients' (1997: 158). In terms of the news coverage being provided, Garrett maintains that too many general assignment reporters lost sight of the microbe by framing the story as a sociopolitical crisis, while too many science reporters relied on 'test-tube stories' which failed to acknowledge the social dimensions of the epidemic. Stories at both ends of the continuum, she points out, were a disservice to the public.

In Britain, similar forms of prejudice affected the quality of news reporting in detrimental ways. Karpf's (1988) research suggests that the British media's coverage of AIDS has been a 'textbook moral panic', thereby playing 'a prominent role in freighting the subject with fear' (Karpf 1988: 141). Much of this coverage, she argues, contributed to a sense of panic surrounding the condition, with the prevailing image of patients being that of social outcasts. Journalists, she maintains, were inclined to see themselves as 'simply doing their jobs' when they 'reported on police reluctance to bring AIDS sufferers into court ("Don't put them in spitting distance – magistrate"), or firemen and ambulance crews' announcement that they'd no longer give mouth-to-mouth resuscitation of injured men' (1988: 144). Further examples, she adds, included those news items reporting the 'theatre cleaners' threat not to clean up after a gay production, or the banning of gay men from a working-men's club ("BANNED! AIDS fear club ousts gay couple"), or quoted those saying that having a haircut could now be lethal' (1988: 144). The media clamour about a 'gay plague' had as its subtext, she argues, 'an outcry about a plague of gays'. The recurrent message was that AIDS was contagious and that homosexuality was a contagion. The country's gay people, just like those living in the US, had become scapegoats. 'The infection's origins and means of propagation excites repugnance, moral and physical, at promiscuous male homosexuality', declared an editorial leader in *The Times* newspaper. 'Many members of the public are tempted to see in AIDS some sort of retribution for a questionable style of life' (cited in Karpf 1988: 143).

Wellings (1988), in her analysis of national newspaper coverage of the epidemic between 1983 and 1985 in Britain, also provides a number of important insights. Evidently prior to the intervention of the National Union of Journalists (NUJ) in 1984, AIDS was virtually synonymous with the term 'gay plague' in press accounts. This description, she argues, surfaced as frequently in the broadsheet or 'quality' press (' "gay plague" may lead to blood ban on homosexuals', *Daily Telegraph*, 2 May 1983; 'gay plague sets off panic', *Observer*, 26 June 1983) as it did in the tabloid or 'popular' press ('watchdogs in "gay plague" blood probe', *Sun*, 2 May 1983; 'alert over "gay plague" ', *Daily Mirror*, 2 May 1983). By the middle of 1984, though, British medical journals were witnessing an intense debate over whether AIDS was a 'new' disease or, alternatively, an 'old' one endemic in Central Africa which had not been properly recognized before. Significant differences were becoming apparent between the African cases of AIDS, and those in Europe and North America. As Wellings (1988) notes, chief among them were, first, differences in sex distribution (the ratio of African men to women with AIDS was almost even) and, second, the African patients did not appear to fall into the particular 'risk groups' previously understood to be most pertinent (see also Epstein 1996).

Debates over the possibility that AIDS originated in Central Africa served, in turn, to render problematic the previously taken-for-granted assumption that it originated among homosexuals. The broadsheet newspapers, in reporting on these debates, called into question the familiar contention that AIDS was a 'gay plague'. In the words of a *Guardian* article, published on 31 October 1984, 'Homosexuals have always objected to AIDS being called the gay plague. The evidence now shows that they are right' (cited in Wellings 1988: 85). Nevertheless, according to Wellings, the tabloid press did not report on the possible heterosexual and African origins of AIDS until 1985. The *Daily Mirror* was the first to take the story up, reporting on 5 March of that year: 'AIDS is believed to have struck first in Africa and spread rapidly.' It then goes on to state: 'from Africa, the disease leapt the Atlantic to the Caribbean and then to the United States. Infected Americans brought it to England and Europe and now the "gay plague" has spread right around the world to Australia' (cited in Wellings 1988: 86). Paradoxically, as Wellings observes, the phrase 'gay plague' is nevertheless still being retained in the report to refer to AIDS. Moreover, those living with the condition continued to be stigmatized, seemingly unworthy of empathy, compassion or concerted action.

Pride and prejudice

'Sometimes I have a terrible feeling that I am dying not from the virus, but from being untouchable.' These words, spoken by a person with AIDS quoted in *The Guardian*, movingly testified to the lived misery of the condition. And yet it seemed during the mid-1980s that for every journalist committed to ensuring that their reporting had a sufficiently strong basis in the apparent facts of the matter, many more were simply content to fan the flames of controversy. News items often consisted of little more than the latest scandalous claim about who or what was 'responsible' for AIDS – with many reporters deciding, as noted above, to blame the victims themselves. An item from the *Sun* newspaper (1 February 1985), headlined 'Gay Plague Kills Priest', is illustrative in this regard:

> Grim-faced ministers emerged from a Cabinet meeting, fearful that the killer plague AIDS will spark violence on the streets of Britain. The prospect of bloodshed as terrified citizens make 'reprisal' attacks on homosexuals and drug addicts is now seen as a real threat. Some gays are expected to retaliate by spreading the virus to the rest of the community through 'revenge sex' with bisexuals.
>
> (cited in Karpf 1988: 143)

This type of scare-mongering, typically accompanied by exhortations to government ministers to intervene to protect the 'moral order', amplified the larger panic. Just as phrases such as 'AIDS suspect' or 'AIDS carrier' effectively criminalized those living with the condition, press reports claiming that a sexual 'hygiene drive' was necessary conflated homosexuality with disease. 'I'd Shoot My Son If He Had AIDS, Says Vicar!' read a *Sun* (14 October 1985) headline. Not to be outdone, the *Daily Star* openly called for the creation of 'leper-like colonies' where those with the condition could be exiled, on the grounds that 'the human race is under threat' (cited in Watney 1997).

Such forms of news coverage, as Altman (1986: 12) remarks, helped to engender a sense of AIDS as a curse of the 'other', something striking groups already singled out for misfortune (see also Murray 1991; Redman 1991). This division was particularly pronounced with respect to the widespread tendency in media reports to differentiate between 'innocent' and 'guilty' victims of HIV/AIDS. Markedly different styles of reporting tended to be adopted, depending on whether the person who had contracted the syndrome was gay or heterosexual. While the deaths of individual gay men were usually only summarily mentioned in the British press, Wellings (1988) observes, the deaths of heterosexuals outside the 'high risk' groups were prominently featured. Generally speaking, she argues, 'the fleshing out of personal biographies is a way that would allow readers to empathize with those concerned has been far more common in relation to cases of AIDS involving non-homosexuals' (Wellings 1988: 87). Moreover, she adds, 'the human faces put to the clinical cases have tended to be those of heterosexuals who have inadvertently caught the disease via a route which was other than sexual' (1988: 87).

Not surprisingly, by this logic, haemophiliacs and recipients of blood transfusions, as well as children and elderly people, received far more sympathetic coverage than those deemed to have been engaging in 'illicit', 'unnatural' or 'morally unacceptable' practices. The experiences of 'innocent' victims, part of the 'blame-free population', were recurrently set in sharp opposition to those of the allegedly culpable minority regarded as 'deserving' the disease. Citing a news report in the *People* (now defunct) newspaper, Watney (1988: 54) shows how it 'colludes with an underlying racism and misogyny which contrasts the "respectable" housewife with AIDS with the "infested" African prostitute'. Other newspapers were simply content to describe AIDS as 'an Act of God', and as such a form of retribution on those who had sinned. In the words of one *Sunday Telegraph* (10 February 1985) columnist:

> Is it not time that the bishops brought God into the act, since one suspects that religious fanatics – condemned by homosexuals as ignorant bigots – who talk about the wrath of God may know more about the cause of the disease, and its cure, than at present do all the scientists working together.
>
> (cited in Eldridge 1999: 112)

Compassion among journalists was all too frequently reserved for the 'innocent victims', while more often than not those categorized in 'high risk' groups remained faceless statistics, effectively depersonalized.

Significantly, however, the ideological underpinnings of this polarization of people with AIDS into the 'innocent' and the 'guilty' began to slowly shift with the death of actor Rock Hudson in 1985. Longstanding rumours that the actor had contracted the syndrome were confirmed in July of that year, with much of the media attention initially centring on a photograph of Hudson kissing fellow actor Linda Evans only months earlier. Hudson, who in the words of one press account was the 'last of the traditional square-jawed, romantic leading men', would give a face to the faceless plague. On 2 October, a spokesperson announced that the actor was dead, a revelation that dramatically increased the scope of AIDS coverage in general. In Britain, one of the *Daily Mail*'s (3 October 1985) writers seized the opportunity to extend a homophobic argument:

> America is gripped with fear, loathing and hysteria over the relentless increase of the unexplained killer disease AIDS [. . .] When Rock

> Hudson admitted he had AIDS, the gay community exploited that fact with near joy. At last they had a public figure, a hero who was one of them. The biggest name in AIDS. The reality has been that it focussed attention on AIDS and also on the causes of it. The gay parades are over. So too is public tolerance of a society that paraded its sexual deviation and demanded rights. The public is now demanding to live disease-free with the prime carriers in isolation.
>
> (cited in Watney and Gupta 1990: 16)

Meanwhile in the US, the number of AIDS stories in newspapers, according to Alwood (1996: 234), 'jumped by an estimated 270 percent in the final months of 1985'. Other factors were also involved, of course, not least President Ronald Reagan's first public acknowledgement of the epidemic, by then in its fourth year, the month before Hudson's death (Reagan's silence had not been challenged by the mainstream media). Nevertheless, the reason so many news organizations began to cover AIDS issues systematically, and others were moved to report on the condition for the first time, was due to the actor. Described by *Time* magazine (14 October 1985) as 'perhaps the most famous homosexual in the world', and now 'the most celebrated known victim of AIDS', it observed that his passing 'belatedly focused public attention on the disease that killed him'. Indeed, his illness and death, Treichler (1999: 74) suggests, constituted a 'critical turning point in the evolution of consciousness about AIDS', decisively changing both media and public notions of who could be an AIDS patient. According to Kinsella (1989: 144–5), the 'simple fact is, newsmakers, from the executive producer at the major network to the assignment editor on the metropolitan daily in Des Moines, were for the first time touched in a direct, personal way by the epidemic'. In the words of an editorial in the *USA Today* newspaper on the day of Hudson's death: 'Many of us are realizing that AIDS is not a "gay plague," but everybody's problem' (cited in Kinsella 1989: 145).

In Britain, the passing of the 'quintessential all-American male' was also front-page news, with much of tabloid coverage still framed in homophobic terms. The report in the *Daily Mail* stated: 'He died a living skeleton – and so ashamed.' The *Sun*'s account was headlined: 'The Hunk Who Lived a Lie: He loved only Mum', while the *Star* declared: 'Hollywood made the legend, Rock Hudson lived the lie.' Hudson's death, as Karpf (1988: 145) observes, 'anchored the illness ever more firmly to homosexuality in the media's mind, and the coverage so closely linked his death with his dissimulation of his homosexuality that it almost seemed as if he'd died from deception rather than AIDS'. This line of argument resonates with Sturken's (1997) assessment, as she argues that Hudson's disclosure 'forced into public view a deeply conflicted image of the person with AIDS – the embodiment of American manhood now contaminated, wasting away' (Sturken 1997: 151). His death, in her view, exacerbated 'the crisis of American masculinity', as he was 'perceived as perpetrating a betrayal [. . .] and a fraud upon the American public, to which he had represented good, safe (heterosexual) manhood' (1997: 151; see also Beynon 2002). At the same time, as Goldstein (1991) points out, with Hudson's death US television news executives 'abruptly discovered the "human-interest" aspect of AIDS' (Goldstein 1991: 26). Moreover, he adds, they 'realized that uncertainties about who might be at risk could draw a huge audience' (1991: 26).

Challenges to the status quo were gaining precious ground, with gay and lesbian activists proving to be particularly effective at having their voices of dissent heard through alternative media outlets and grassroots organizations. The by now familiar prejudices of those mainstream news organizations reluctant to take HIV/AIDS seriously were at last slowly – far too slowly – beginning to crumble and occasionally give way to more sensitive forms of reporting. The pressure for change was increasingly being directed at newsrooms from a diverse array of vantage points. Organized protest marches for gay and lesbian rights, not least over the right to adequate health care, sparked considerable media attention, thereby helping to contest official definitions of the crisis. In the US, activist groups such as GLAAD (the Gay and Lesbian Alliance Against Defamation) and ACT UP (the AIDS Coalition to Unleash Power) developed a number of media-savvy strategies to get their messages onto mainstream news agendas. At stake was the aim of articulating their demands for improved government initiatives to combat the epidemic, as well as to call into question the imperatives of medical and scientific research. Of particular concern in the case of the latter was the pressing need to speed up the approval of new drugs, to further democratize access to such treatments, and to ensure that people with AIDS were cared for in a humane manner.

ACT UP's demonstrations were designed to shock, and they usually succeeded. Members distributed free condoms and safe-sex pamphlets to people outside of religious institutions following prayer services. Protesters demanding lower prices for HIV/AIDS medicines occupied drug company offices. Others chained themselves to a banister at the New York Stock Exchange, while same-sex 'kiss-ins' were staged at political conventions. One such ACT UP demonstration, taking place on the first day of classes at Harvard Medical School in 1988, involved activists dressed in hospital gowns, blindfolds and chains spraying fake blood on the pavement. They then raised their voices to chant: 'We're here to show defiance / for what Harvard calls "good science"!' (Epstein 1996: 1). An outline of a mock

AIDS 101 class was handed out to the university's students, where discussion topics included:

> PWA's [People With AIDS] – Human beings or laboratory Rats?
>
> AZT – Why does it consume 90 percent of all research when it's highly toxic and is not a cure?
>
> Harvard-run clinical trials – Are subjects genuine volunteers, or are they coerced?
>
> Medical elitism – Is the pursuit of elegant science leading to the destruction of our community?
>
> (Epstein 1996: 1)

Also by 1988, the AIDS memorial quilt was beginning to attract regular media attention around the globe. The quilt – made up of thousands of small, multicoloured panels each bearing the name of someone who had died from the condition – provided a powerful media image of the ongoing tragedy in stark, human terms. More than producing a 'collective body count', the quilt became a deeply political act of remembrance as it was taken around the world. One panel, for example, reads: 'I have decorated this banner to honor my brother. Our parents did not want his name used publicly. The omission of his name represents the fear of oppression that AIDS victims and their families feel' (cited in Sturken 1997: 187). Later that same year, television journalist Max Robinson, the first full-time black anchor of a network newscast in the US, died of AIDS. Having constantly fought for improved employment conditions for ethnic minorities in broadcasting, his earlier characterization of the media as 'a crooked mirror' through which 'white America views itself' had assumed a new resonance where HIV/AIDS was concerned.

Nevertheless, despite these and related events, quantitative research suggests that the overall amount of news coverage of AIDS appearing worldwide was on the wane as the decade came to a close (see Kinsella 1989; Lupton 1994; Tulloch and Lupton 1997; Miller et al. 1998). Newspaper editors, like their counterparts in radio and television news, were much less inclined to carry stories about the epidemic, and on those occasions when items were published they were usually relegated to the back pages. 'Seven years after AIDS first appeared on the front page of an American newspaper,' Alwood (1996: 237) observes, 'there was still no sign of a cure for the disease, and news decision-makers considered stories about it too routine and too depressing.' A 1990 survey of gay and lesbian journalists conducted by the American Society of Newspaper Editors found that a significant number of respondents felt that while their employers were 'tolerant' of them, they reported 'widespread homophobia in the newsroom' (*Time* magazine, 16 April 1990). They also pointed to certain 'distortions' in their newspapers' coverage of 'gay-related issues', not least HIV/AIDS, and found management uninterested in hearing their ideas about how best to improve it.

Meanwhile, as the death rate grew steadily worse, public interest in the epidemic appeared to be in an alarming state of decline. By 1991, with the media preoccupied with the 'Gulf War' in the Middle East, members of the activist group ACT UP decided the time was right for an intervention (the group by now had dozens of chapters in the US, as well as others in London, Berlin and Paris). They were angered by the US government's apparent willingness to spend a reported $1 billion per day to conduct the war, when a total of $5 billion had been spent on AIDS research in the ten years of the epidemic. On 22 January, members of the New York chapter dramatically interrupted the live broadcast of the *CBS Evening News*, and the Public Broadcasting System's *MacNeil-Lehrer NewsHour*, in what they called their 'Day of Desperation'. In the case of the CBS newscast, three protesters burst into the studio just as the newscast began, shouting 'Fight AIDS, not Arabs!' Newsreader Dan Rather, understandably startled, promptly halted his introduction and declared 'We're going to go to a commercial now.' Several seconds of blacked-screen silence then ensued before Rather reappeared to restart the newscast with an apology to viewers for the 'rude people in the studio' who, by now, had been 'ejected' (and later arrested). Meanwhile, seven protesters in the *MacNeil-Lehrer NewsHour* studio were struggling to chain themselves to the newsdesk and studio equipment, as well as to the anchorperson, Robert MacNeil. Their actions failed to appear onscreen, however, as viewers were presented with co-anchor Jim Lehrer reporting from Washington instead. MacNeil later explained to other reporters that a 'group of non-violent protestors' had 'complained that we and the media are spending too much time and attention on the war in the Middle East which they say will never kill as many people as are dying of AIDS'. Attempts to storm the news studios of ABC and NBC were foiled by security guards.

To this day in the US, however, the greatest surge in the volume of news coverage about HIV/AIDS took place in the autumn of that year. On 7 November 1991, NBA star Earvin 'Magic' Johnson announced his retirement from professional basketball, a decision taken after he tested HIV-positive. Speaking at the press conference, he declared: 'We think, well, only gay people can get it – it's not going to happen to me. And here I am saying that it can happen to anyone, even me, Magic Johnson.' Precisely how he contracted the virus was not revealed; indeed he bravely stated that it did not matter. The next day though, while a guest on a television talk show, he told the host: 'I'm far from being homosexual. You know that. Everybody else who's close to me understands that.' The studio audience, according to Crimp (1993: 258), 'went berserk, cheering wildly for several minutes', evidently feeling vindicated following this reassurance that they had not been 'duped into hero-worshipping' someone who was gay. Shortly thereafter, in an interview with a sports magazine, Magic Johnson would add: 'I confess that after I arrived in L.A. in 1979, I did my best to accommodate as many women as I could – most of them through unprotected sex.' The player's use of the word 'accommodate' in this context was judged by many to be inappropriate, shifting as it did blame for the virus onto the women involved while rendering their respective risks invisible (see Rowe 1997). The ensuing news coverage did further public understanding of the condition, however. As Maddox (1991: 103) observed at the time, 'it may be cold comfort for Johnson that his case should make such an important contribution to public education'.

It is important to note that Johnson's announcement continues to be regarded by many commentators in the US media as the single most significant factor affecting public awareness and concern about HIV/AIDS to date. The athlete, as *Time* magazine maintained, was ideally placed to speak on the condition, especially with respect to impoverished minority communities and the young. 'Though blacks represent only 12% of the nation's population,' one *Time* reporter pointed out, 'they account for 25% of the AIDS patients: more than half the women with AIDS in the U.S. are African American' (*Time* magazine, 18 November 1991). Many of the most well-intended AIDS-prevention programmes, he added, 'have failed to speak the language of the groups that are most at risk'. In the days following Johnson's revelation it became apparent that for many young people, remembering where they were when they heard the news was akin to the impact of hearing about the Kennedy assassination for their parents. It similarly became clear that the ensuing news coverage of Johnson's new life helped to recast public images of the condition. 'For years we tried to get the media to distinguish between HIV and AIDS;' observes Crimp (1993: 261), 'they finally found it necessary to comply in order to reassure Magic's fans that he was infected with HIV but did not have AIDS.' Moreover, he writes, 'we asked the media for images of people with HIV disease living normal, productive lives [and they] gave us Magic playing the All-Star game' (Crimp 1993: 261). News accounts dealing with the issue of whether Johnson might infect other players with his blood, such as in the event of a collision, helped to focus public attention on the means by which the virus is transmitted. Johnson also became an outspoken critic of President George Bush's failure to mobilize sufficient resources to fund medical and scientific research into HIV/AIDS and to adequately treat those living with the condition. Having agreed to serve on the President's National Commission on AIDS, he would eventually resign in protest several months later over the issue of funding.

HIV/AIDS in a new century

'Whatever happened to AIDS?' asked the headline of a news item in the *Washington Post* newspaper in September 1991. 'After ten years, AIDS no longer captures as much attention,' the journalist observes. 'Yet the virus has spread to nearly every county in the US. So far, 118,411 are dead. One million more are estimated to be infected' (cited in Alwood 1996: 239). Two months later, as discussed above, Magic Johnson's announcement would focus intense media scrutiny on the condition. Also in 1991, the US Federal Food and Drug Administration recommended that Haitians be banned from donating blood, a decision which Haitian-Americans (still stigmatized in the press as a 'high risk group' with 'tainted' blood) took to the streets to protest (see Dubois 1996). Lost on many in the news media was the crucial distinction between 'at risk' groups and 'at risk' behaviours and practices, their attention to the condition having by then proved to be both fickle and fleeting. Indeed, it seemed to be the view of many journalists that HIV/AIDS had become yesterday's news.

In the first decade of the new century, some journalists have been posing the same question. Over the course of 2001, the twentieth anniversary of the first early indications of the impending epidemic, mainstream news organizations typically offered retrospective assessments. Commentators pointed to a range of startling advances, as well as bitter setbacks, in what has been a collective struggle to come to terms with HIV/AIDS ever since. Still, in the case of the US, the apathy of those journalists who believe that the crisis is over is difficult to overcome, despite the types of figures readily provided by health care researchers. Since 1981, more than 750,000 cases of AIDS have been identified in the country, and almost half a million people have died from it. About 800,000 to 900,000 of its people are estimated to be currently living with HIV/AIDS, with some 40,000 new cases every year (about one-quarter of whom are women). There remains a disproportionate spread of the disease among heterosexual black women, who make up 7 per cent of the US's population yet evidently accounted for 16 per cent of all new AIDS diagnoses in 1999. 'By comparison,' Sack (2001) points out, 'black men made up 35 percent, white men 27 percent, Latino men 14 percent, and white and Latino women were each 4 percent.' HIV/AIDS, tragically, is anything but old news, but to report on it adequately necessarily means confronting the very sorts of issues, such as homophobia, racism and poverty, which mainstream news organizations increasingly shy away from covering. In general, however, the tone of articles covering HIV/AIDS has improved over the years; many of the attitudes expressed in the extracts cited earlier in this chapter would seem anachronistic in present-day editions of the same newspapers.

In comparative terms with the US, infection rates per capita are less severe in the UK, but are still a far more serious concern than typically acknowledged by journalists. HIV has infected almost 50,000 people since the first cases began to appear in 1981. Figures compiled by the Public Health Laboratory Service and the Scottish Centre for Infection and Environmental Health offer cumulative data to the end of March 2001. Specifically, the figures show:

> there have been 44,988 reported cases of HIV in the last two decades, of which 14,038 people have died. Of the HIV total, 25,806 are believed to have resulted from sex between men, 11,667 from sex between men and women, 3,695 from intravenous drug use, 1,351 from blood-clotting factor used predominantly to treat haemophilia, 751 from mother to baby transfer, and 314 from blood transfusion and tissue transfer (1,404 remain undetermined). Of the 44,946 case reports in which sex was stated, 36,398 were men, 8,548 women.
>
> (Garfield 2001)

The year 2000 saw 3617 newly reported cases of HIV in the UK, more than in any year since the epidemic began (and an increase of 16 per cent from 1999). Research by the Terrence Higgins Trust suggests that the number of people living with HIV in the UK is expected to increase by 50 per cent in the five years to 2006 (*Guardian*, 28 November 2001). Nevertheless, health officials point out, AIDS is widely seen as 'yesterday's disease'. Journalists, like other members of the public, are perceived to have become largely indifferent to the dangers posed by the ongoing crisis. According to a 2000 report published by the Department of Health, as many as 10,000 people in the UK have HIV but are unaware of their condition. Some activists in HIV/AIDS awareness projects believe that many people at risk today are too young to remember the major government public health campaigns from the 1980s. Pertinent information these days, they argue, is more likely to reach them via popular culture, such as via HIV storylines in films and television dramas, than through official channels. In the case of *EastEnders*, for example, the character Mark Fowler has been HIV-positive since 1991, allowing the programme to address a range of issues, such as safe sex and prejudice from the community. Evidently, the character's illness 'prompted more people to have HIV tests than the entire expensive government public health campaign of the late 80s and early 90s', while one 1999 survey found that 'most teenagers had got their information about Aids from *EastEnders*' (*The Guardian*, 21 June 2001).

Even where real progress in getting the message across is being made, however, entrenched poverty – and the ways it is implicated in access to health care, unemployment, substance abuse, sexually transmitted diseases and inadequate schools – continues to place limits on what can be achieved. Successes in reducing HIV/AIDS incidence rates can be all too quickly erased by complacency about the underlying structural inequalities endemic to even the wealthiest of countries. Turning to the developing world, however, the implications these inequalities engender are even more painfully apparent. AIDS is now poised to become the world's leading cause of death. According to a report by the Joint United Nations Programme on HIV/AIDS, 5.3 million new cases were diagnosed in 2000. Since 1981, almost 22 million people are thought to have died from AIDS, about 9 million of whom being women. By the end of 1999, some 13.2 million HIV-negative children under the age of 15 had lost either both parents or their mothers to AIDS – 12.1 million of them in Sub-Saharan Africa. As Rosenberg (2001) writes in *The New York Times*:

> Someday, we may look back on the year 2001 with nostalgia for a time when AIDS was merely a health catastrophe. Soon, AIDS in Africa will be doing more than killing millions every year. It will destroy what there is of Africa's economy and cause further instability and, perhaps, war. In the year 2010, the country of South Africa will be almost one-fifth poorer than it would have been had AIDS never existed.
>
> (Rosenberg 2001)

Following South Africa, the country with the next greatest number of people living with the condition is India, with almost 4 million people infected with HIV. Unsafe drug-injecting practices represent the most common way the disease is being spread. The epidemic arrived relatively late in Asia, but according to UN figures published in 2001, the region accounts for about 15 per cent of all infections worldwide. The rate of HIV infection is rising faster in Central Asia and Eastern Europe (where Ukraine has the region's highest prevalence) than anywhere else in the world (Altman 2001b).

In China, where the epidemic has been officially recognized only since 2001, at least 600,000 of its people are estimated to be HIV-positive – a figure believed by some to be set to increase tenfold by 2006 (see also

Watanabe 2001). Factors making HIV a challenge to control in places like China, according to Rosenthal (2001), include: 'ignorance, denial, discrimination, weak laws and a rural health system that is expensive, corrupt and virtually bankrupt.' While the growing use of condoms in some developing nations is helping, infection rates for sex workers are a particularly worrisome concern (and male clients, in turn, placing their partners at risk). Such is also true with regard to intravenous drug use, and the spread of HIV through blood transfusions. In the case of the latter, the problem is largely due to some hospitals finding blood from official blood banks so expensive that they rely instead on poor blood sellers. Methods used to collect the blood are often unsanitary, the blood itself is not always tested, and the sellers themselves – usually living in abject poverty – are typically not provided with medical check ups. While the epidemic continues to be ignored in some countries (for some governments it is regarded as an embarrassment to be covered up), in others – such as Thailand and Cambodia – large-scale AIDS-prevention programmes are underway. Control and prevention strategies based on public education, given a chance to work, are already proving to be effective in helping to counter the spread of the disease. These programmes, however, remain the exception to the rule.

By adopting this kind of global perspective, it quickly becomes apparent that not only is the AIDS epidemic rapidly escalating, it is threatening to quickly overwhelm efforts to control it. At the time of writing, the scientific breakthrough leading to a vaccine remains over the horizon, although some of the more optimistic voices in the scientific community are confidently predicting success within the next few years. In the meantime, antiviral drugs have some effect on the virus, but are intended for short-term use due to their toxicity, which can produce debilitating side-effects, and they are also very expensive. Nevertheless, in countries where such treatments are available, health care researchers are increasingly pointing to what seems to be a resurgence of unsafe sex among some gay men, with some reports suggesting 'barebacking' or unprotected anal sex is becoming more popular. 'If unprotected sex now equals longtime therapy,' observes one HIV/AIDs activist, 'all of a sudden a lot of people who were scared are taking a whole lot of risk.' For some, it seems, 'exhortations about H.I.V. are like cautions about cigarettes, illicit drugs or driving too fast, warnings from "old people" about distant dangers' (*The New York Times*, 19 August 2001). Still, while those living in wealthier counties may be able to think of HIV/AIDS as being a chronic condition made manageable with daily pill-taking regimens, for those living in poorer countries it remains a fatal one. For governments of impoverished countries, patents on AIDS drugs can be so costly as to put treatment out of the reach of their fragile health care systems. As a result, they argue, there often appears to be little choice but to consider allowing generic copies of the drugs to be made without the permission of the companies owning the patent. Some pharmaceutical companies, alarmed at the prospect, have responded by sharply reducing the price of their products. From their point of view, however, their profits can be justified due to the expenses associated with investing in the research and development of new medicines. Whether or not this is indeed the case is a question for investigative journalists to pursue, yet too few seem willing to do so.

It is clear, as Altman (1986: 11) wrote at an earlier stage of the crisis, that 'the way in which AIDS has been perceived, conceptualized, imagined, researched and financed makes this the most political of diseases'. For the journalists covering HIV/AIDS over the years, many would acknowledge that the limitations of their reporting are much clearer with the benefit of hindsight. It is important to bear in mind that they too did not know what they were dealing with at first, so their own, personal fears were understandable. That said, of course, these fears do not even begin to excuse the use of the homophobic or racist language to characterize the condition. To the extent that certain journalists have been complicit in stigmatizing people with HIV/AIDS, they have violated their most basic responsibilities to their fellow citizens.

Today, as the crisis continues to unfold, critical research into how the news media cover the attendant issues remains as vital as ever. The concept of 'moral panic' may seem almost anachronistic to some, but several of the factors which it helps to elucidate are very much alive in certain places around the world. It remains important, of course, to ensure that use of the concept is sensitive to the materiality of different situational contexts. That is to say, after Watney (1988, 1997), it is crucial that the concept be handled in a manner which recognizes different orders or degrees of what is likely to be an array of multiple, interrelated moral panics at any one instance. 'We are not living through a distinct, coherent "moral panic" concerning AIDS,' he writes, 'a panic with a linear narrative and the prospect of closure' (Watney 1988: 60). Rather, he adds, 'we are witnessing the ideological manoeuvres which unconsciously "make sense" of this accidental triangulation of disease, sexuality, and homophobia' (1988: 60). Dickinson (1990) similarly points to the limitations of the moral panic concept, identifying several areas of difficulty in its application to HIV/AIDS. His main concerns include the need to recognize the historical specificity of any one moral panic, particularly the extent to which it re-inflects aspects of previous ones, and the diversity of ideological representations across different genres of media texts. It follows that the ways in which media portrayals of HIV/AIDS are structured, at least in part, by existing media discourses of medicine, health and illness, warrant closer scrutiny. Consequently, Dickinson (1990: 27) argues, the concept is 'of use only insofar as it offers helpful guidelines for exploring the complexities of the communication process' (see also Miller et al. 1998; Thompson 1998; Ungar 2001; Critcher, in press).

It is this sense of the moral panic concept as a useful starting point for developing a heuristic framework that has been adopted in this chapter. While I have focused on the contours of news coverage in Britain and the US, with a particular emphasis on the early years when the primary definitions of the crisis were being established, it goes without saying that alternative approaches would have been equally valuable. Not only has important work been undertaken on the news coverage of HIV/AIDS in other national contexts, of course, but also other researchers have taken as their foci the fluidly complex conditions under which this coverage has been both produced and consumed.

Related research has examined the dynamics of news production, for example, showing how the day-to-day imperatives of making HIV/AIDS 'newsworthy' occurs in relation to a diverse array of institutional constraints. Questions surrounding the negotiation of 'news values' were significant in this regard, as were the restrictions encountered by some journalists over what could be reported, and how. Efforts to report on the transmission of the disease, for example, were sometimes hampered by the refusal of many news organizations (as a matter of editorial policy) to allow words such as 'semen' or 'vaginal fluid', let alone 'anus', to be used (see Kinsella 1989; Alwood 1996). Research into how journalists frame HIV/AIDS similarly helps to disclose the strategies by which different news actors, some of whom were stakeholders with vested interests (and axes to grind), seek to advance their definition of the situation (see Murray 1991; Lupton 1994; Miller et al. 1998; Eldridge 1999). Related studies of public information campaigns have also provided relevant insights (see Atkinson and Middlehurst 1995; Carter 1997; Hurley 2001), as have those centring the media activities of 'moral entrepreneurs', policy-makers, health care workers, scientists and activists (see Patton 1985; Fauvel 1989; Nelkin 1995; Epstein 1996; Garrett 1997). Other researchers have unravelled generalized assumptions regarding 'public perceptions' about HIV/AIDS so as to examine how members of different audiences make sense of news reports, especially where issues of risk are concerned, as part of everyday life (see Tulloch and Lupton 1997; Miller et al. 1998). In this context, moreover, audience members' engagements with news coverage similarly need to be situated in relation to representations of HIV/AIDS as they are depicted in other media genres, such as current affairs, documentary or drama (see Dickinson 1990; Goldstein 1991; Bird 1996; Sturken 1997; Treichler 1999; Long 2000).

To close this chapter on an encouraging note, we may recall the optimism expressed by UN Secretary General, Kofi Annan, speaking in June 2001 at a three-day summit in New York devoted to the epidemic. 'This year we have seen a turning point,' he declared. 'Aids can no longer do its deadly work in the dark. The world has started to wake up. We have seen it happen in the media and public opinion – led by doctors and social workers, by activists and economists and, above all, by people living with the disease' (*The Guardian*, 26 June 2001). It is indeed the people living with HIV/AIDS, in my view, who need to have their voices heard much more clearly by journalists striving to do justice to the cacophony of claims and counter-claims about the realities of the disease. Forms of journalism which attend to their lived experiences, their stories of pain and sorrow but also hope, will help to ensure that the human consequences of this condition are more widely acknowledged and, I want to believe, respected.

Further reading

Alwood, E. (1996) *Straight News: Gays, Lesbians and the News Media*. New York: Columbia University Press.

Epstein, S. (1996) *Impure Science: AIDS, Activism and the Politics of Knowledge*. Berkeley, CA: University of California Press.

Kinsella, J. (1989) *Covering the Plague: AIDS and the American Media*. New Brunswick, NJ: Rutgers University Press.

Lupton, D. (1994) *Moral Threats and Dangerous Desires*. London: Taylor & Francis.

Miller, D., Kitzinger, J., Williams, K. and Beharrell, P. (1998) *The Circuit of Mass Communication: Media Strategies, Representation and Audience Reception in the AIDS Crisis*. London: Sage.

Sturken, M. (1997) *Tangled Memories*. Berkeley, CA: University of California Press.

Treichler, P.A. (1999) *How to Have Theory in an Epidemic*. Durham, NC: Duke University Press.

Tulloch, J. and Lupton, D. (1997) *Television, AIDS and Risk*. St Leonards, NSW: Allen and Unwin.

Watney, S. (1997) *Policing Desire*, 3rd edn. London: Cassell.

FOOD SCARES: MAD COWS AND GM FOOD

Food comes first, then morals.

(Bertolt Brecht, playwright and poet)

The line of dialogue above, borrowed from the second act of Brecht's 1928 play *The Threepenny Opera* (*Die Dreigroschenoper*), strikes a rather curious resonance in relation to current debates about food scares and the media. Public perceptions of risk, trust and uncertainty where food is concerned are inextricably entwined with moral issues. If, for some, our consumption of food has never been such a hazardous activity as it is today, others insist that thanks to scientific innovations it has never been safer. Regardless of whether one considers the distance between food and morals to be widening or narrowing, however, many would likely agree that the most pressing concern is that millions of people on this planet are going hungry each day. Even for those fortunate enough to have the benefit of a sufficiently nourishing diet, though, everyday decisions about what to eat – or, more to the point, what not to eat – appear to be assuming a far greater moral significance than before. It almost goes without saying that the range of choices implied by these seemingly mundane, taken-for-granted decisions are as political as they are personal.

Media discourses concerned with food – encompassing everything from campaigning consumer journalism to celebrity chef cooking shows, advertisements, recipes in magazines, and so forth – continue to proliferate at a remarkable rate. Here it is significant to note the extent to which issues around food safety, in particular, are being made the subject of scrutiny across this mediascape. This is hardly surprising when one considers that in a country such as Britain about 100,000 food poisoning cases are reported each year, a mere fraction of those thought to be actually taking place. According to figures from the government's Food Standards Agency, there were 65,209 laboratory confirmed cases of food poisoning in 2000. The five most common bacterial **pathogens** causing the majority of cases were **salmonella**, campylobacter, listeria moncytogenes, ***E. coli*** and clostridium perfringens (FSA press release, 23 August 2001). Meanwhile in the US, according to the Centers for Disease Control and Prevention, every year about 5000 deaths, 325,000 hospitalizations and 76 million illnesses are caused by food poisoning. Figures for Australia, from the Australia New Zealand Food Authority, indicate that on any one day there are 5700 to 8600 new cases of food poisoning, approximately 1 in 30 of which is officially recorded.

Statistics like these ones beg the question, as Winter (2001) aptly poses it: 'Why, in an age of technologies that protect food, is food poisoning at least as common as it was a half-century ago?' Possible answers are varied. Improved methods of detection are clearly significant. At the same time, intensive farming practices mean that many animals continue to be raised and slaughtered under conditions that invite bacterial defilement. Variations in people's lifestyles are also frequently cited as contributory factors. Researchers point to changes in shopping habits, where fewer trips to the shop mean food is stored for longer periods, sometimes improperly so. People also appear to be eating more fresh fruit and vegetables without cooking them first, thereby increasing the chance of infection via bacteria or viruses. The popularity of so-called 'convenience' foods (which, despite their name, often require more careful preparation than is recognized) is also a factor. Moreover, the sheer range of foods becoming available is outstripping governmental efforts to inspect them, leading to the emergence of new strains of micro-organisms from faraway places. The global transport of foods, coupled with the need to extend their shelf-life, means that the use of chemical treatments, irradiation and genetic engineering also may be giving rise to further hazards (see also Powell and Leiss 1997; Adam 1998; Ford 2000).

Consumers, it seems, are often the first to test a particular food's safety. 'We are the canaries in the coal mines,' one professor of public health argues. 'The moment someone gets sick, we say, "Don't eat that food." It's a miracle that the system doesn't break down more' (cited in Winter 2001). As a result, food scares have become a regular occurrence in many countries, with new food-related threats emerging all of the time. 'Today,' according to Brody (2001), 'almost no food is exempt from possibly harboring a micro-organism or microbial toxin that can make people sick and, in some cases, kill them.' Moreover, she adds, while 'millions of consumers worry about pesticide residues and additives in foods, the real hazards lie in microbial contamination and improper food handling, especially in the home'. In this context, the media are often credited with playing a vital role in alerting people to the need to be ever vigilant where food is concerned. Others maintain that an enhanced public awareness of food risks means that instances of food contamination or poisoning that might have gone unreported in the past are now much more likely to be front-page news, whether they deserve to be or not. Still others, taking up the latter theme, blame the media for scare-mongering, for frightening a gullible public into a panicked state over what are usually inestimably minute risks. Indeed, as Reilly and Miller (1997: 234) observe, the media are recurrently 'seen as irresponsible and sensationalist, either by uncritically allowing the nation's health to be damaged by the food industry or causing undue alarm by publicising the views of non-experts, pseudo-scientists and politically motivated pressure groups'.

In order to further elucidate several key issues surrounding food risks, this chapter takes as its focus news media coverage of food scares. Particular attention is devoted to news reporting about two ongoing crises: first, the identification of BSE or 'mad cow disease' in Britain's cattle and the ensuing 'beef crisis'; and, second, public fears about genetically modified (GM) crops and foods. In the first instance, though, the anatomy of a food scare, in general, will be discussed in the next section. 'Eating', as Caplan (2000: 187) observes, 'has become a risky business.'

Food scares in the media

Another day, another food scare. With almost frightening regularity, it seems, news reports identify a problem, real or potential, with one type of food or another being consumed somewhere around the world. What may have appeared to be safe enough to eat yesterday, a 'food expert' may be declaring to be unduly risky today. In the absence of adequate advice about food quality and risk standards, the ensuing confusion can threaten to erode public trust and confidence. 'People are being bombarded with so many conflicting, changing messages,' Dr Wendy Doyle of the British Dietetic Foundation observes. 'I feel the public are getting a little sceptical, and feel, rightly or wrongly, that they might as well eat what they like and not listen any more, which is not really what we want' (cited in *The Guardian*, 22 June 2001). If, as Doyle suggests, some members of the public react by deciding to ignore expert pronouncements, others will frequently display alarm, sometimes even panic, in their response to media reports.

Any perceived threat to public health, needless to say, is likely to be regarded as potentially newsworthy by journalists. Food scares make for good headlines, arguably ones which boost sales, and the mere hint of controversy is likely to attract the attention of a wide array of pertinent stakeholders, each with their own media agendas to pursue. Efforts to discern the impact of news reports can point to what is arguably a growing number of 'kitchen panics' taking place. A brief list over recent years, composed primarily by drawing on newspaper sources, includes the following:

- 1988: British Conservative MP Edwina Currie, a Junior Health Minister, announces that most of the country's eggs are infected with salmonella, resulting in the slaughter of more than a million hens.
- 1989: an outbreak of listeria frightens British consumers away from such foods as pâte, cook-chill foods and soft cheeses.
- 1989: in the US, a study suggests that the chemical Alar, a 'plant growth regulator' used in the apple industry, is a 'potent carcinogen' when processed. A frightened public scorns apples and apple products.
- 1990: traces of benzene, a carcinogen, are found in a well-known brand of bottled spring water. The company recalls its entire inventory, temporarily halting production.
- 1993: apple juice is once again in the media spotlight, as high levels of a natural mould, patulin, are found by British researchers.
- 1993: *E. coli* in minced beef used by a fast-food chain in its hamburgers results in 732 victims across four states in the US. Almost 200 people were hospitalized, 55 of whom developed kidney failure, while 4 children died.
- 1996: in the face of previous denials, British government officials acknowledge the possibility of a link between bovine spongiform encephalopathy (BSE) or 'mad cow disease' and the fatal new variant of Creutzfeldt–Jakob disease (vCJD) in humans.
- 1996: over 9500 people in Japan, largely schoolchildren, were stricken with *E. coli*, 12 of whom died. In November, 400 people fell ill – 21 of whom died – from the same bacterium in Scotland.
- 1998: scientist Dr Arpad Pusztai announces to journalists that rats suffered ill health when fed potato genetically engineered to produce an **insecticide**.
- 1999: a faecal contamination of South Australian orange juice affected almost 500 people.
- 1999: the maker of a popular soft drink recalls some 2.5 million bottles of its product after dozens of its customers are hospitalized in Belgium. Soon after, Belgian meat and eggs are reportedly contaminated by dioxins due to tainted pork and poultry feed.
- 2000: in Japan, more than 14,000 people become ill after drinking contaminated milk.
- 2000: the British government reports that 23 per cent of pigs taken for

slaughter are infected with salmonella. Alarm is raised over existence of toxins in British salmon.

- 2000: it is reported that in India 3000 people are hospitalized after eating contaminated rice.
- 2001: 'unacceptable' levels of a carcinogenic chemical are found in 22 out of 100 soy and oyster sauces tested by the British Food Standards Agency.
- 2001: following a case of infant botulism, a British baby food manufacturer withdraws from sale batches of formula milk substitute.
- 2001: researchers for a BBC documentary find that more than two-thirds of fresh supermarket chickens are infected with a food poisoning bacteria, namely the campylobacter jejuni strain.

These food scares, among others which might have been listed, raise important questions about how media discourses concerned with food influence public perceptions of health and risk. Some of the above scares had a basis in the devastating consequences for the people involved, while others following investigation were eventually proved to be entirely unfounded. In either case, however, the complex array of factors shaping the 'food scare' phenomenon clearly warrant much greater attention from researchers than they have typically received to date.

One of the few attempts to distinguish the principal features of a food scare is that of Beardsworth and Keil (1997) in their sociological inquiry into the human food system. A food scare, they suggest, may be characterized as 'an acute outbreak of collective nutritional anxiety which can seize hold of public awareness and can give rise to significant short- and long-term consequences' (Beardsworth and Keil 1997: 163). In their view, a 'typical food scare' is recognizable as such because it exhibits a fairly consistent pattern, consisting of a sequence of steps:

> 1 An initial 'equilibrium' state exists in which the public are largely unaware of, or are unconcerned about, a potential food risk factor.
> 2 The public are initially sensitized to a novel potential food risk factor.
> 3 Public concern builds up as the risk factor becomes a focus of interest and concern within the various arenas of public debate.
> 4 Public response to the novel risk factor begins, often consisting of the avoidance of the suspect food item. (This response may be an 'exaggerated' one, apparently not in proportion to the 'actual' risk.)
> 5 Public concern gradually fades as attention switches away from the issue in question and a new 'equilibrium' state establishes itself. However, chronic low-level anxiety may persist, and can give rise to a resurgence of the issue at a later date.
>
> (Beardsworth and Keil 1997: 163)

From Beardsworth and Keil's perspective, then, it is vital to explore the dynamics engendering these surges of public concern. While observing that the notion of 'moral panic' (discussed in Chapter 6) is potentially useful in this regard, an even more suitable way forward in their view is to examine 'news spirals'. That is to say, where moral panics revolve around the ways in which a 'deviant' group may be characterized as posing a threat to social order, the concept of a news spiral allows researchers to discern a broader range of factors contributing to public anxiety. Significant here is the 'feedback loop' by which public reactions to media portrayals of an issue are themselves re-inflected by journalists as news in subsequent stages of the coverage. 'The reporting of audience reactions, Beardsworth and Keil (1997: 164) maintain, 'itself increases public awareness of the issue and increases the level of sensitization, thereby closing the feedback loop and allowing a spiralling level of anxiety to build up.' This process of anxiety 'amplification', they suggest, is capable of producing acute surges in collective fear or apprehension, at least until the issue eventually becomes 'stale' in news terms.

Of particular interest in these relatively intense, albeit frequently short-lived, news spirals is the extent to which scientific discourses about the risks associated with food are contributing to public anxiety. Scientific discourses, Beardsworth and Keil point out, are by no means necessarily reassuring ones for the public, particularly in the absence of satisfactory or unambiguous responses to exigent questions. 'Questions which appear to require no more than a recourse to suitable facts or data to settle them once and for all can turn out to be highly intractable,' they observe, 'with little prospect of clear solutions in the medium or even the long term' (1997: 159–60). Scientific pronouncements about possible risks – the accredited voices of nutritionists, dieticians, physicians, microbiologists, toxicologists and physiologists, among others, being especially salient in this context – routinely appear to be mired in debate and controversy. Constant disputes between experts, far from being reported on by journalists as simply part of the rough and tumble of how science is conducted, 'can convey to the public a sense of confusion, indecision and even incompetence in the face of what are seen as serious threats to well-being' (1997: 167). This is the case, Beardsworth and Keil maintain, despite the unprecedented success innovations in food production, processing, storage and transportation have had in raising public expectations concerning food quality and safety. 'These high expectations themselves,' they argue, 'make a direct contribution to the public's susceptibility to food scares' (1997: 167; see also Meek 2001).

News coverage of ongoing disputes over food risks evidently plays a decisive role in their politicization. Journalists are increasingly prying open for public scrutiny the processes informing official decision-making, thereby helping to create – at least in principle – the conditions for voices otherwise excluded to be heard. 'Once issues of food safety', Beardsworth and Keil (1997: 171) write, 'have escaped the control of a closed, oligarchical policy community and truly entered into the public domain, the official voice becomes only one voice among many, each presenting its own competing account.' In the absence of scientific certainties, officials are increasingly hard pressed to sustain public tolerance, let alone generate confidence or reassurance, where possible food risks are concerned. These kinds of tensions, as the next section will illustrate, have become all too apparent in the 'mad cow disease' crisis that continues to unfold in one country after the next around the world.

Mad cows, angry citizens

More than any other food scare in modern Britain, the 'mad cow disease' crisis appears to have provoked the most intensely felt fears, suspicions and misgivings about the quality of the country's food supply. The first cases of BSE, or bovine spongiform encephalopathy, were detected in cattle in the mid-1980s. Over a decade would pass before the then Conservative government would finally acknowledge what many members of the public had long come to suspect, namely a probable link between BSE and vCJD, the fatal new variant of Creutzfeldt–Jakob disease in humans. People eating meat from cows infected with BSE, the government admitted, were theoretically at risk of developing the untreatable brain disease. Almost overnight the beef market collapsed, with devastating social and economic effects. Looking at the crisis today, it is apparent that lasting damage has been done to public trust in government ministers and officials, and the scientists who advise them. In the words of Parliament's BSE Inquiry report, published in October 2000: 'When on 20 March 1996 the Government announced that BSE had probably been transmitted to humans, the public felt that they had been betrayed. Confidence in government pronouncements about risk was a further casualty of BSE' (Phillips 2000).

How did such a state of affairs come about? Evidently the first signs of what would become known as BSE surfaced in the herd of a West Sussex farmer in December 1984. Confronted with a cow behaving abnormally (head tremors and a loss of co-ordination), he had contacted a veterinary doctor for assistance. The vet, puzzled by the symptoms, consulted scientists at the Central Veterinary Laboratory (the cow, called 'cow 133', died in February 1985). In time the laboratory would formally identify the mysterious affliction, the symptoms being described in a clinical report as indicative of a 'novel progressive spongiform encephalopathy in cattle'. BSE was deemed to be a **prion** disease, prions being mutated proteins which can no longer be broken down by the body. Their brain-riddling effect, frequently described as transforming the brain until it resembles a sponge with large holes, was identifiable only through autopsy (prions could not be detected in blood or tissue tests at the time). Typical symptoms include the loss of bodily control and dementia. One Wiltshire farmer described the progress of the disease as follows: the cow 'begins losing weight, walks unsteady and sometimes a bit sideways. It stares; it begins shaking, lowing. It's nervous, upset. It can go berserk. It may charge' (cited in Moeller 1999: 72). In light of this kind of behaviour in infected cattle, BSE would eventually acquire the 'mad cow disease' tag in media reports.

Almost from the outset, scientists speculated that the infected cattle had been fed ruminant offal which, in turn, had been contaminated. To clarify, offal is made up of those portions from rendered cattle or sheep, such as entrails, tissue and hooves, not being processed for human consumption. The contamination, it was surmised, had most likely been derived from offal removed from sheep suffering with scrapie (also a degenerative brain disease). It is worth noting in this context that this practice of converting herbivore cows into carnivores by feeding them ground-up bits of animals (or 'protein supplements' as they are sometimes called) is intended to make them grow at a faster rate than ordinary. For some critics of this practice, it constitutes a form of 'industrial cannibalism', one that is simply 'unnatural'. Taking up the latter theme, one *Guardian* journalist would later ask: 'Did it not occur to anyone that feeding animal protein to what nature evolved to be a herbivorous species might be dangerous?' (cited in Brookes and Holbrook 1998: 177).

British government officials were first informed about BSE in June 1987 when the Chief Veterinary Officer told the Ministry of Agriculture, Fisheries and Food (MAFF) about the new disease. The official line, adopted from the outset ostensibly on the basis of the 'best scientific advice', was that there was no risk of the disease being passed to humans. The first news story in the mainstream media, according to Eldridge (1999), appeared in the *Sunday Telegraph* (25 October 1987) under the headline: 'Incurable Disease Wiping Out Dairy Cows'. Written by the newspaper's agriculture correspondent, the account points out that the source of the disease was unknown, nor was it apparent what caused it or how it might be cured. Television news evidently picked up the story a month later, pointing out that although MAFF officials baffled and worried about the disease, 'there's no need to panic' (cited in Eldridge 1999: 115). No further television news stories were forthcoming until the following year. In April 1988, the government indicated that legislation would be instated to render BSE a notifiable disease. Officials had established the Southwood Working Party to investigate concerns that BSE was entering the food chain. Shortly thereafter, a ban on the use of meat and bone meal feed was introduced (although, crucially for other countries, it was not extended to exports), together with a slaughter policy for all animals showing BSE symptoms. Evidence was published in the *Veterinary Record* in October of that year suggesting that BSE could be transmitted to mice via an inoculation of diseased matter from cows. At year's end, the Agriculture Minister, John MacGregor, reaffirmed once again that 'there is no evidence that the disease can be transmitted to humans.'

In July 1989, the export of British cattle born before July 1988, together with the offspring of affected animals, was banned by the European Union. It had become apparent that the disease would continue to escalate so long as ever greater numbers of cows ate infected feed and then had their infected remains fed to yet more cows. In November 1989, a ban was imposed on the use of high-risk offal (namely, the brain, spinal cord and certain organs such as the spleen) for human consumption. It was feared that hundreds of thousands of contaminated beef carcasses had entered the human food chain. If this was so, then it meant that millions of people in Britain were unwittingly eating products – such as beefburgers, cheap mince and pies – made from meat infected with BSE. In November 1989, the government banned the use of the most prion-rich cattle tissues in all human food. Nevertheless, in the months that followed, a number of official assurances were made, most notably from the Ministry of Health, that British beef posed absolutely no risk to humans. Public scepticism continued to grow, however. In May 1990, a farmyard cat, called Max, was reported to be dying of a BSE-like disease. Some scientists hypothesized that BSE had crossed the species barrier naturally (incidentally, it was this development which, according to Powell (2001: 221), prompted the first coverage of BSE in US newspapers). Professor of Microbiology Richard Lacey made the first public call for all infected herds in Britain to be slaughtered. Other scientists, however, begged to differ. Responding to the press reaction to the dead cat, a *New Scientist* editorial expressed annoyance:

> One cat dies, 'A Scientist' speaks, six million cows are supposed to hit the knacker's yard and you can't even turn the bovine bodies into feline fodder. Or, to put it another way, hysteria rules . . . the media are at it again.
>
> (cited in Gregory and Miller 1998: 175)

The absence of evidence, for some scientists, was evidence of absence. Presumably with this type of logic in mind, the government's Chief Medical Officer once again sought to assure members of the public that beef remained safe to eat.

Later that same month, Agriculture Minister John Gummer decided to intervene in order to try to reduce the chances of a full-fledged panic. Facing the television cameras with his 4-year-old daughter, Cordelia, he proceeded to encourage her to eat a beefburger, stating: 'It's delicious. I have no worries about eating beefburgers. There is no cause for concern.' These images, the subject of much satiric comment in the press, would in time return to haunt the minister and his government. In September, researchers announced the results of laboratory experiments where BSE appeared to have been transmitted to a pig. Still insisting that there were no implications for human health, government officials nevertheless stepped up the ban on 'specified bovine offals' in all animal feed. By 1992, it was calculated that in Britain three cows in every thousand were infected with the disease. The government's Spongiform Encephalopathy Advisory Committee (SEAC) advised in May that the existing measures taken to safeguard human and animal health were adequate. Sir Kenneth Calman, Chief Medical Officer, reiterated the government's assurance that British beef was safe to consume on 11 March 1993.

Two years later, this assertion had begun to crumble. People were beginning to die from a strange and distressing new form of dementia. The disease was named as a 'new variant' of Creutzfeldt–Jakob disease, a longstanding illness to which it bears some resemblance. A progressive neurological disorder, standard CJD is rare and sporadic, with patients almost always aged over 55 years. The new variant of CJD is a distinctively different disease, one far more likely to strike young people. Symptoms include bouts of depression, mood swings, numbness, followed by hallucinations, uncontrolled body movements, memory loss and dementia. It progressively cripples the brain, leaving patients completely immobile and mute until their central nervous system collapses. The process generally takes up to 18 months. It is, to be blunt, an absolutely horrific way to die. The first known variant CJD fatality was 19-year-old Stephen Churchill, who passed away on 21 May 1995. His death was followed by that of another teenager that year, and several others – teenagers, as well as people in their twenties and thirties – over the ensuing months. All were decades younger than the ages of those usually afflicted with traditional CJD.

Further media coverage of these seemingly isolated instances of variant CJD began to highlight potential connections being drawn by the scientists themselves. By November 1995 it was becoming increasingly apparent that infected tissue was still entering the human food chain. MAFF scientists conducting spot checks on abattoirs found that some were ignoring the ban placed on specified bovine offal. Dressed carcasses were sometimes found to contain pieces of spinal cord. The possible implications were being clearly spelt out by journalists. In the words of an item in *The Guardian*:

> People who ate infected meat in the late 1980's may not show signs of the disease until the next millennium. Then, if the doomsday scenarios proved right, biblical numbers could suffer loss of co-ordination, intellect and personality – just like the wobbly mad cows seen on TV.
>
> (*The Guardian*, 8 December 1995)

Scientists themselves were actively involved in the media fray. Sir Bernard Tomlinson, a leading brain disease expert, reportedly told BBC Radio 4's *You and Yours* consumer programme that he had stopped eating meat pies or beef liver and 'at the moment' would not eat a beefburger 'under any circumstances' (cited in Powell 2001: 222). A serious cause for concern among some scientists was the use by food companies of 'mechanically recovered meat' (MRM) in their products. MRM, scientists hypothesized, carried a serious risk of infection because much of it is made up of tissue residue situated close to the spinal column. Following this kind of advice, MAFF announced in December a ban on the addition of MRM to human food. Schools began banning beef from cafeteria menus. Still, Prime Minister John Major told the House of Commons that month: 'There is currently no scientific evidence that BSE can be transmitted to humans or that eating beef causes CJD.'

Government 'spin' (or 'information management') was decisively unravelled, once and for all, on 20 March 1996. The Health Secretary, Stephen Dorrell, stunned members of the House of Commons by conceding that there was a 'probable link' between BSE and vCJD. He stated that 'a committee of scientists set up to advise the government on the issue had linked an unusual outbreak of the human disorder, Creutzfeldt–Jakob disease, to exposure to the cattle disease.' Ten cases of the new form of CJD, he admitted, had been diagnosed in human patients. Dorrell's earlier insistence that 'there was no conceivable risk' in eating beef was no longer tenable. The inconceivable had suddenly become officially conceivable. A news story in that morning's edition of the *Daily Mirror* had stated the connection succinctly: headlined 'Official – Mad Cow Can Kill You', it had arguably succeeded in forcing the government's reversal. As Adam (1998: 164) points out, 'we cannot even begin to guess how the entire fiasco would have developed if the *Daily Mirror* had not leaked the story' (see also Adam 2000). In any case, by finally recognizing that people were falling victim to a disease most likely caused by exposure to BSE infected meat, the government invited a popular backlash. Members of the public were left to conclude, as Macnaghten and Urry (1998) observe, that a cynical cover-up had been at work behind what appeared to be deliberately misleading statements. Even for those who accepted 'the idea that governments know from science the true nature of the risks associated with such diseases as BSE', the fear remained that 'they are not telling us these facts which they really know' (Macnaghten and Urry 1998: 260; see also Jasanoff 1997; Durant 1998b; Ratzan 1998; Rhodes 1998; Baker 2001).

Among the factors leading to Dorrell's announcement was the pressure being brought to bear by news coverage of vCJD victims. An item in the *Daily Express* newspaper, titled 'The Mad Cow Deceit: Fourth Teenager is Killed by CJD', is indicative. Its opening paragraph pinpoints the link that Dorrell had previously denied:

> Creutzfeld-Jacob Disease used to take decades to kill. Sufferers tended to be over 55. But CJD is changing. Now it kills the young. Can it be a coincidence that, in the last 10 years, we've eaten 1.5 million BSE-infected animals?
>
> (*Daily Express*, 18 February 1996)

The news item proceeds to report on the death of 20-year-old Peter Hall of CJD, the fourth such teenage victim. A first year university student, he had enjoyed excellent health before the disease made its presence known. In relating its account of his decline, the news item explicitly seeks to challenge what it calls 'the Government's "no evidence" orthodoxy'. In a direct appeal to the reader, the item states: 'You will be shocked (but not surprised) to learn that the Government is still not investigating their deaths.' Three months to the day of Dorrell's statement to the House of Commons, Frances Hall, the mother of Peter, would intervene. Speaking on BBC Television's *Newsnight* (20 June 1996) programme, she declared:

> Our son Peter was ill for more than a year. During that time I wrote to request that someone from the government would come and sit with me at his bedside and see what this devastating illness was doing to our strong, handsome young man. No one came . . . Will the government now accept that the scientific advice it chose to follow, namely, that there was no conceivable risk from eating British beef, was wrong? Are the experts still the same? Is the government still being selective on the advice that it takes on behalf of the nation or is it now willing to err on the side of caution? These past months have been, and continue to be, a living nightmare for my family. We have been unable to come to terms

with Peter's horrible death because we know that if BSE had been treated with sufficient caution he and many others would not have suffered this terrible illness.

(*Newsnight*, cited in Cottle 2000: 29–30)

Precisely why so many vCJD sufferers have been young people quickly became the subject of intense speculation among journalists and their sources. Some scientists called upon for answers postulated that a disproportionate amount of contaminated meat had found its way into school meals prior to the bovine offal ban. Others would later hypothesize that young people are more vulnerable to infection because of changes in their teeth, or because they are more prone to tonsillitis or sore throats. By this theory, inflammation of the gums, or in the throat area, becomes a route of exposure. Fears were also raised that it might be possible to pass vCJD through contaminated blood products, donor tissue or even via surgical or dental instruments (evidently the infectious prions can survive standard sterilization processes) during medical procedures.

In the days following Dorrell's announcement to Parliament, several long-standing critics of the government's no-risk to humans stance – some of whom having been described as 'prophets of doom' or as 'charlatans', 'sensationalists' or worse – suddenly found they enjoyed far greater media credibility than before. Some of them seized the opportunity to call for the end of intensive farming methods, while others had their predictions of vCJD casualties (some calculations putting the figure at 10 million by 2010) given extensive coverage. Newspapers such as the *Daily Mirror* were drawing distinctions between 'Whitehall scientists' and a 'growing band of independent experts' critical of the government's preferred position (Brookes 2000). Still, some newspapers were reluctant to criticize the government, even likening the opposition party's leading critic, Harriet Harman, to a 'mad cow' for being a 'panic monger'. The act of criticism itself was enough for *The Sun* (26 March 1996) to level the charge of promoting hysteria:

HAS THE WORLD GONE MAD?. . .

It's not the cows who are mad. It's us humans.

Over the past six days, we've put our brains out to grass.

Hysteria has taken over from common sense as we contemplate destroying the beef and dairy industry – and four million cattle – because of yet another food scare . . .

And that's all it is. A SCARE . . .

The panic which has gripped the British public has been whipped up

> by those with an axe to grind. LABOUR'S Harriet Harman has behaved disgracefully. She twists public concern into political point-scoring.
>
> In the Commons yesterday she gave another *wild-eyed* performance. She blamed Tory de-regulation. But farming is one of our most regulated, controlled industries.
>
> Has mad Harriet ever spoken to a farmer . . .
>
> So hands off our cows. Eat British beef . . .
>
> And stop talking bullocks
>
> (*The Sun*, cited in Brookes and Holbrook 1998: 180)

Far more significant than Conservative or Labour Party point-scoring, however, were the ways in which a potential public health crisis was being recast as a 'beef crisis'. As Adam's (1998, 2000) analyses suggest, by framing the BSE–vCJD link in strictly economic terms vis-à-vis the beef industry, several implications follow:

> 1 Citizens' concerns about a potential health hazard are sidelined.
> 2 The Government can legitimately concentrate exclusively on the material task of rescuing a dire economic situation.
> 3 The entire issue can be dealt with from the firm basis of facts since (it is wrongly assumed) economics is not afflicted like science by the malaise of uncertainty.
> 4 People need not concern themselves with questions about the industrial methods of food production.
> 5 The UK Government, aided and abetted by the press – the tabloids more so than the broadsheets – can put their full efforts behind the effective externalisation of blame.
>
> (Adam 1998: 185)

Most journalists accepted and reproduced this official reformulation of the crisis, she argues, but not because trickery or deception were involved. Rather, 'such a reframing of the problem was a means by which to bring this rogue issue back within the fold of the familiar news world of reportable statements and events, describable disasters, and quantifiable economic facts and figures' (Adam 1998: 186). The significance of BSE, this logic dictated, was to be measured solely in relation to safeguarding the financial interests of a key industry for the British economy.

Despite the concerted efforts of officials to define the issue this way, however, the possible BSE–vCJD link became such an intense site of public anxiety that it could not be effectively managed or contained within these terms entirely. Compounding their difficulties in sustaining the 'beef is safe' message, McDonald's promptly declared that it would stop selling British beef products in its restaurants (Dutch beef was to be used instead), a decision soon followed by the Burger King and Wimpy's chains. Public confidence was further undermined, as Moeller (1999: 73) observes, when 'a reporter showed up at former agriculture minister John Gummer's house during the first week of the crisis with a hamburger in hand . . . and Gummer turned it down'. Reilly and Miller (1997), in their discussion of the reasons why BSE did not receive sustained media coverage until Dorrell's statement, quote an all too telling remark made by one of the broadsheet newspaper journalists they interviewed:

> It's logical really. Newspapers demand new information, new angles, controversy what have you. I couldn't get BSE in all the time. They lost interest in the subject because nothing was happening. Of course that was the whole point, nothing was happening to destroy this thing, but in newspaper terms I wouldn't be given the space to say that every day or every week. At the same time a few of us were seen as being slightly OTT [over the top] on the subject, a bit nutty. I don't think people really believed that there was a real danger from beef – there were no dead people (at that time) so, in a sense, although I was given a lot of scope, what had to happen for the full-scale go-ahead of a major story was dead people. Well, we've got them now.
>
> (cited in Reilly and Miller 1997: 244–5)

By the end of March 1996, the European Union had imposed a worldwide ban on all exports of British beef. The government responded the very next day by announcing that tighter BSE controls would be implemented, including a 30-month slaughter scheme so as to stop cows older than that age from entering the human food or animal feed chain. Speaking at an EU summit in Florence, Prime Minister Major declared a 'beef war', confidently predicting (wrongly as it turned out) that a legal challenge to the EU ban would see it lifted by the end of the year.

The election of a Labour Party government in May 1997 brought with it a change of approach. It was widely perceived that the crisis had helped to seal the fate of the Conservative government among voters. Throughout the election campaign, serious questions had been raised about the culture of secrecy and blame-passing within the Major administration and the civil service. During its last months in office, the Conservative government had introduced a selective cull of cattle regarded as being most at risk from BSE in an attempt to convince the EU to lift its export ban. It was criticized as too little, too late. Labour ministers announced new controls for scrapie in June, ordering that all sheep suspected of having the disease be slaughtered. In December, the sale of 'beef on the bone' was banned, triggering another flurry of media interest. Brookes (2000: 204) cites several of the pertinent headlines:

> WHAT IS TRUTH ABOUT BEEF? Latest fiasco confuses public once again
>
> (*Daily Express*, 4 December 1997)
>
> BEEF: WHAT ARE WE TO BELIEVE NOW? Chaos for shoppers as meat on the bone is banned
>
> (*Daily Mail*, 4 December 1997)
>
> WHAT THE HELL CAN WE EAT? Chaos as oxtails, T-bone steaks and beef ribs face chop
>
> (*Mirror*, 4 December 1997)

An editorial published in *The Sun* the following day, also cited by Brookes (2000: 206), places the blame squarely on the Labour government:

> WHEN BSE first burst on to the menu, *The Sun* asked a simple question: *Has the world gone bloody mad?*. . .
>
> If you eat a T-bone steak, you're 1,000 times more likely to choke to death than die of mad cow disease.
>
> *And yet the government has caused a panic by banning beef on the bone*. Of course, the Agriculture Minister had to act quickly . . .
>
> But now we've finally seen how flimsy the case is against beef on the bone, the Government would have been better off telling people the facts and letting them make up their own minds.
>
> (*The Sun*, 5 December 1997)

When the 'beef on the bone' ban was eventually lifted two years later, several newspapers proclaimed a victory for 'common sense' over official assessments of risk. Meanwhile, the new Labour government set up a BSE Inquiry in December 1997 to 'reveal the events and decisions which led to the spread of CJD and BSE'. By March 1998, public hearings into the BSE crisis were underway. In November, following the introduction of a new cattle-tracing system, EU agricultural ministers (with the exception of France's) agreed to lift the ban on exports of British beef, which eventually took place in August 1999. Early in 2000, a mother with vCJD gave birth to a baby girl who, tests showed, had also contracted the disease.

In October 2000, the BSE Inquiry submitted its 16-volume, 4000-page report examining the causes of the BSE crisis and its handling by the government. Evidence for the report had been gathered from a wide range of sources since 1998, including public hearings where, among others, families of vCJD victims were asked to testify. Its opening words capture something of the public anxieties associated with the disease:

> BSE has caused a harrowing fatal disease for humans. As we sign this Report the number of people dead and thought to by dying stands at over 80, most of them young. They and their families have suffered terribly. Families all over the UK have been left wondering whether the same fate awaits them.
>
> (Phillips 2000)

Promptly dubbed the Phillips report by journalists, after Lord Phillips who led the two-year inquiry, it strongly criticized former ministers and officials in the then Conservative government for failing to co-ordinate an adequate response to the crisis. Instead the possible risks for humans were consistently played down as the government chose to adhere to its 'campaign of reassurance' in the hope that consumer panic could be allayed and the loss of beef exports avoided. 'At the heart of the BSE story', the report states, 'lie questions of how to handle a hazard – a known hazard to cattle and an unknown hazard to humans.' It credits the government for taking 'sensible' measures to address both hazards, but points out that 'they were not always timely nor adequately implemented and enforced'. In short, ministers had placed the demands of producers, namely the farmers, over and above the interests of consumers who needed to be sufficiently informed about potential risks. Ministers similarly sought to limit the financial expenditure necessary to turn policy into practice, even to the point of having no contingency plans for dealing with a vCJD epidemic. Although the government 'did not lie to the public about BSE', the report adds, it was 'preoccupied with preventing an alarmist over-reaction'. As a result, 'the possibility of a risk to humans was not communicated to the public or to those whose job it was to implement and enforce the precautionary measures.'

So far, 4.5 million cattle have been slaughtered in the UK under the 30-month scheme introduced in 1996 (once again, intended to ensure cows older than that age do not enter the food chain). A further 70,000 have been destroyed under the selective culling scheme introduced in early 1997, with several thousand more killed, or waiting to be killed, under the BSE offspring cull introduced in 1999. Diagnostic tests performed on the animals' carcasses have consistently indicated that a much smaller proportion of them were actually infected. Today, the wholesale culling of herds is no longer taking place, due to new research on prions which has enabled assays to be developed which can test animals for BSE while they are still alive. Sadly for many farmers just beginning to recover from the misery of the crisis, the outbreak of foot-and-mouth-disease set them back again (see Brassley 2001).

At the time of writing, over 100 people are believed to have died as a result of eating BSE-infected meat in Britain. Precisely how large the scale of vCJD will be continues to be a matter of considerable dispute among scientists. Given that no one knows for certain how long the disease remains undetected in humans before symptoms emerge, a variety of different scenarios are being envisaged. Some scientists are convinced that vCJD has already reached its peak. Others insist that the disease's lengthy incubation period (some believing it may be as long as 30 or 40 years) means that thousands, possibly hundreds of thousands, more people may be affected in the future. Recent work has reportedly demonstrated that animals with different genetic backgrounds have varying incubation times. As a result, DeFrancesco (2001: 22) suggests, 'an estimate based on the reported cases, all of which had the same genetic background, may be optimistic and that what has appeared to date might only be the tip of the iceberg'. One shared viewpoint is that no one is sure how many people are harbouring the disease, the prions lingering in their blood and tissue. Another, needless to say, is that scientists still have much to learn about the link between BSE and vCJD, particularly with respect to the precise manner of infection (while some scientists, it is worth bearing in mind, continue to refute the claim that any such link exists). Significantly, in the same month that the Phillips report was published, doctors announced that vCJD had claimed as a victim a 74-year-old man, the oldest known victim. The ensuing press coverage sparked further fears about the potential size of the possible human epidemic to come.

Identified cases of BSE in the UK still far outnumber those of any other country, although they are now in decline. Research suggests that the epidemic peaked in 1992–93. In several European countries, however, the disease is on the increase. In France, for example, the crisis erupted in 2000 with hundreds of affected cattle identified (the first cases also appeared in Germany and Spain in November of that year). Soon after, the first vCJD deaths were reported. A French parliamentary committee released a scathing report in May 2001 concerning the government's handling of the crisis over the previous ten years. Successive agricultural ministers were accused of failing to respond adequately, each of them preferring to try to appease a powerful agricultural sector instead. According to the report, agricultural officials had 'repeatedly sought to prevent or delay the adoption of precautionary measures – that later proved necessary for health safety – on grounds that there was no scientific basis for them'. Similarly subject to harsh criticism, not surprisingly, was Britain. On the basis of the report's findings, the British government 'had to accept "major responsibility" for exporting the disease to Europe by "shamelessly" authorizing exports of its meat and bone meal when it had already concluded that such feed components were an important element in transmitting the disease' (Daley 2001).

In the US, at the time of writing, there have been no confirmed cases of BSE or vCJD. A ban has been in place on the import of all beef products from the UK since 1989 (and in Canada since 1990). Shortly thereafter a further ban was imposed on the feeding of any ruminant-based by-product to cattle, although 100 per cent compliance has yet to be achieved according to recent governmental research. New rules have also been adopted to prohibit anyone who had lived in Britain from donating blood, a ban later extended to include people from other European countries as well. So far, though, there appears to be little indication, judging from opinion surveys, that very many members of the public are concerned about BSE in the US food supply. The same data suggest that most are confident that governmental authorities, as well as food producers, will avoid the mistakes of their European counterparts (*The New York Times*, 26 January 2001).

One brief moment of panic, however, was sparked by the 'Dangerous Food' episode of the *Oprah Winfrey Show* (broadcast 16 April 1996) which discussed the BSE crisis in Britain, and raised questions about whether or not something similar might happen in the US. Indicative here is a statement made by Oprah in response to a claim made by one of her guests: 'That has just stopped me cold from eating another burger' (cited in Brookes 2000: 199). In the days following the broadcast, beef sales temporally plummeted in some parts of the country. Several Texas cattle ranchers proceeded to sue Winfrey, the producers and distributors of her talk show, and one of her guests. Alleging that the beef market suffered substantial losses following the broadcast, they argued that the Texas False Disparagement of Perishable Food Products Act had been violated. The courts rejected their case, along with the subsequent appeal. Oprah's reaction, widely quoted in news accounts at the time, was: 'Free speech not only lives, it rocks.'

The first cases of vCJD outside of Europe were reported on 9 February 2001. Two patients were diagnosed with the illness in Thailand, a country which had banned beef imports from Britain five years earlier. In June of that year, the three largest international agencies for health and agriculture – the World Health Organization, Food and Agriculture Organizations and World Animal Health Organization – called for all countries to assess their risk of BSE. Governments were urged to adopt precautionary measures to prevent the global spread of BSE. On 10 September 2001, BSE reached Japan in what was regarded to be the first 'native-born' case outside of

Europe (cases of BSE in other countries outside of Europe, such as Canada, Oman and the Falkland Islands / Islas Malvinas, were due to imported British cows). Officials in Japan believed that the cow, a 5-year-old Holstein, had been infected by contaminated British feed imports. Some encouraging news to emerge, however, concerns the potential for new drug treatments for vCJD. At the time of writing, UK-based clinical trials were being set up for quinacrine, an anti-malaria drug. The trials have been prompted by the temporary remission of a British female patient, believed to be suffering from vCJD, being treated in the US. She received a combination of quinacrine with chloropromazine, an anti-psychotic drug. While it is far too early to make predictions, this treatment has raised hope that a vaccine might be possible one day.

The controversy over genetically modified foods

'Who's afraid of genetically modified foods?' the front cover of *The Economist* magazine (19–25 June 1999) demands to know. Situated to one side of the headline is a potato drawn to resemble a grimacing human face, complete with metal electrodes jutting out (Frankenstein's monster-like) from its chin on both sides. The answer to the question, as reported on the inside pages, is just about everyone.

Genetically modified – GM, for short – foods are proving to be much more than just another passing food scare. The subject of an intensive consumer backlash not seen in Britain since the peak of the BSE or 'mad cow disease' crisis, the media storm has engulfed politicians, scientists, policy-makers, food industry spokespeople, retailers, farmers, protesters and ordinary citizens in a whirlwind of competing claims and counter-claims. In the words of the editorial leader published in the same issue of *The Economist*:

> If the current British furore over genetically modified foods were a crop not a crisis, you can bet Monsanto [the leading biotechnology corporation involved] or its competitors would have patented it. It has many of the traits that genetic engineers prize: it is incredibly fertile, thrives in inhospitable conditions, has tremendous consumer appeal and is easy to cross with other interests to create a hardy new hybrid. Moreover, it seems to resist anything that might kill it, from scientific evidence to official reassurance. Now it seems to be spreading to other parts of Europe, Australia and even America. There, regulators will face the same questions that confront the British government: how should the

public be reassured, and how can the benefits of GM foods be reaped without harm, either to human beings or the environment?

(*The Economist*, 19–25 June 1999)

The editorial line adopted is decidedly pro-business, as might be expected with this magazine, but does recognize voices of dissent. In identifying the proclaimed benefits of GM crops, it acknowledges that 'inadvertently, in hyping the technology as the only answer to everything from pest control to world hunger, the industry has fed the popular view that its products are unsafe, unnecessary and bad for the environment.' Such fears, the editorial argues, are 'largely groundless', with 'much of the public fuss' being 'misplaced'. Nevertheless, it concedes, neither governments or companies can afford to ignore it, particularly when '[f]ood fears and environmental qualms spread more readily than good sense or wise science'. The main issue, from *The Economist*'s perspective, is how best to 'win public support' by 'allaying people's fears about GM food'.

In the early weeks of 1999, the British House of Lords report *Science and Society* (see Chapter 4 of this book) points out, the 'GM food issue' was a 'media storm waiting to happen'. In its view, members of the public had become 'sensitised' to the issue due to a combination of the 'BSE debacle' and the rapid introduction of GM commodity crops into the UK market over the previous two to three years. During the month of February, some newspaper editors began to perceive a widening gap between governmental and industrial policy and practice towards GM crops and food, on the one hand, and public attitudes towards its relative benefits, on the other. 'In this situation,' the report states:

> a single event – particularly one involving scientific dissent from the consensus view that GM foods were safe to eat – was sufficient to trigger a debate in which many newspaper editors, sensing that their readers were generally suspicious about the whole area, decided to campaign against agricultural and food biotechnology.
>
> (Select Committee on Science and Technology, House of Lords 2000)

The event in question that sparked the 'Great GM Food Debate', as the report calls it, was a letter published in *The Guardian* on 12 February 1999. Co-written by an international group of 22 scientists from 13 countries, the letter made public their support for the research findings of Dr Arpad Pusztai (unpublished at the time) which raised the possibility that GM potatoes being fed to laboratory rats were having harmful effects on them.

Pusztai, a geneticist of international reputation, had been summarily suspended and forced to retire from his post with a government-funded research project (at the Aberdeen-based Rowett Research Institute) the previous year after he gave television interviews expressing his concerns over GM foods. On one such occasion, Pusztai had told a reporter for Granada TV's *World in Action* programme that he would not eat GM food himself, and found it 'very, very unfair to use our fellow citizens as guinea pigs'. These concerns had arisen in light of preliminary findings that rats who ate potatoes inserted with a **gene** for a natural insecticide, namely lectin (produced by snowdrops, a type of flower), appeared to incur damage to their organs and to have their immune systems depressed. 'I was totally taken aback,' Pusztai told one journalist. 'I was absolutely confident that I would not find anything. But the longer I spent on the experiments, the more uneasy I became' (cited in *The Independent on Sunday*, 8 March 1999). Describing himself as 'a very enthusiastic supporter of the technology', Pusztai nevertheless added that 'it is too new for us to be absolutely sure that what we are doing is right.' The initial charge made against him by the institute revolved around the allegation that he had presented provisional data in public without the benefit of peer review. Even more serious charges would be made later, all of which were in turn strongly repudiated. The signatories to the letter appearing in *The Guardian* (none of whom, it would be later argued by critics, were noted genetic engineers) maintained that their independent examination of the published data indicated that Pusztai had every right to be concerned. In their view, he should not have been attacked or suspended.

The pressure for a moratorium on GM food, *The Guardian* reported the same day it published the scientists' letter, 'is beginning to look like a tidal wave'. The issue was now the top news story across the media spectrum, with some journalists using the phrase 'Frankenstein foods' to great rhetorical effect. The Labour government faced heated criticism in the House of Commons for not imposing a three-year moratorium on the commercial release of genetically modified crops, even from its own benches. As Labour MP Alan Simpson remarked: 'I think as the Government we have an obligation to identify who frustrated this research? If Dr Pusztai is right, this could be BSE mark two' (cited in *The Guardian*, 12 February 1999). Opposition Conservative MPs, arguably sensing an opportunity to deflect attention away from their party's mishandling of the BSE crisis, challenged the scientific basis for the government's claims. More rigorous testing needed to be done, they argued, a demand echoed by a wide array of individuals and groups, ranging from Greenpeace to the Consumers Association. Despite government efforts to dismiss the criticism as just the latest 'media bandwagon' passing by, it was quickly becoming apparent that the issue – in the words of *The Guardian*'s science editor, Robin McKie – 'had turned into a political nightmare'. In describing the 'unprecedentedly ferocious criticism' being directed at those responsible for making GM foods in Britain, McKie suggested that the products 'now look like the pariahs of the European food industry' (*The Guardian*, 14 February 1999).

Explanations for this 'political nightmare', as might be expected, frequently centred on the role of the news media. Journalists were either praised for finally revealing the truth about GM foods ('Pusztai had been stifled'), or were castigated for scare-mongering ('sloppy science and overblown reporting'). For science editor McKie at *The Guardian*, the origins of the problem could be more accurately attributed to those individuals behind the new technology 'misunderstanding the public's fear of science and failing to realise that consumers become suspicious and vulnerable to fear when they are starved of choice'. In sharp contrast, advocates of genetic engineering insisted that it would provide greater choice, not less, and hence claimed to be perplexed by the apparent failure of critics to appreciate the attendant benefits. Many of these advocates pointed out that crops have been undergoing genetic alterations by random mutation, accident and natural selection for thousands of years. The use of genetic engineering techniques, in their view, would actually make food safer by affording scientists a greater degree of control over this process than is possible with the usual 'hit and miss' techniques. Additional advantages to be gained were said to include:

- GM crops can produce their own pesticides, or be resistant to **herbicides**, thereby leading to less environmental contamination associated with the use of chemicals
- the growers' reliance on fertilizers may be reduced, while the plants are hardier and grow larger, ripen faster and keep longer, thereby making for higher yields
- increased productivity and lower costs for grower make food cheaper to purchase
- they offer consumer better nutrition (e.g., food oils lower in artery-clogging fats) and disease-fighting compounds (e.g., tomatoes rich in cancer-fighting substances or vegetables delivering vaccines)
- promise benefits to developing world as plants thrive in hostile environments
- without GM foods the earth will be unable to feed ever-growing billions of people who inhabit it.

Accurate media coverage, these advocates maintained, would lead to public support for genetic engineering. Such was the case elsewhere, they argued, with seeds created by genetic engineers at such companies as

Monsanto (US), Du Pont's Pioneer Hi-Bred International (US) and Novartis (Swiss) being used throughout North and South America, China and Australia. In the US, as Hopkin (2001) observes, 'an estimated 60 percent of processed foods in supermarkets – from breakfast cereals to soft drinks – contain a GM ingredient, especially soy, corn or canola; some fresh vegetables are genetically altered as well.' Moreover, advocates point out, there are no 'risk-free' foods anyway, with even traditional crops having their risks.

Critics opposed to the new technology took issue with many of the assumptions underlying these claims. In Britain, a powerful anti-GM coalition had already been forming prior to the public debate provoked by the Pusztai letter. Its members encompassed organizations such as Friends of the Earth, Greenpeace (who had been calling for a ban on GM food and crops for the previous decade) and other environmental groups, as well as the Soil Association, farmers and consumer groups. Their objections to the technology included:

- genetic manipulation is against nature, and as such a form of tampering
- the food it produces is dangerous (or not enough research has been conducted to prove that it is safe), with fears that it might cause allergic reactions particularly salient
- its cultivation may lead to unforeseen problems for the environment, in part due to factors such as persistence of insecticides in the soil and evolution of pests that can withstand it
- pollen from engineered crops could render plants in neighbouring fields sterile, while super-weeds (able to resist herbicides and pesticides) might devastate the biodiversity of the countryside
- the technology involved rests with a small number of companies attempting to monopolize crop production (in part by imposing copyright on genes), and to make farmers dependent on them
- adverse effects from GM organisms are likely to be irreversible, and beyond the control of scientists.

Bitterly distrustful of official assurances in the aftermath of BSE, these critics are adamant that insufficient evidence exists to inform a decision to proceed with the commercial planting of modified crops. To do so, they argue, is contrary to the public interest, even in those parts of the world where hunger prevails. Much more food is currently produced around the world than is used, they point out. 'People go hungry not because there is too little food,' one *Guardian* columnist reminded readers, 'but because food and the land on which it grows are concentrated in the hands of the rich and powerful' (*The Guardian*, 13 February 1999).

The furore provoked by the Pusztai letter, a 'media circus' in the eyes of some, reignited the issue of public trust in politicians, scientists and farmers. As Pusztai himself would later state, 'I can say from my experience that if anyone dares to say anything even slightly contra-indicative, they are vilified and totally destroyed' (*The Independent on Sunday*, 8 March 1999). In seeking to better understand how the media shaped public perceptions of the crisis, the Parliamentary Office of Science and Technology (POST), together with the House of Lords Science and Technology Select Committee, commissioned an investigation. Researchers, led by Durant and Lindsey, proceeded to analyse the media coverage – primarily national broadsheet and tabloid daily and weekly newspapers, with additional scrutiny of selected radio and television news and current affairs programmes – between 8 January and 8 June 1999.

The final report, titled *The 'Great GM Food Debate'* (POST 2000a), makes for fascinating reading. Its findings suggest, for example, that although the ensuing debate was 'triggered' by the letter in *The Guardian*, it was not the sole or even necessarily the most significant influence on the controversy. Here the study draws attention to a range of factors involved, including:

- the steady erosion of public confidence in policy-making for food safety following the BSE incidents, particularly after March 1996
- the start of imports of North American commodity crops (soya and maize) into Europe in autumn 1996, which contained unsegregated mixtures of conventional and GM material
- the emergence in the period 1996–98 of a broad, powerful coalition of critics of GM food
- the existence of a small number of prominent individuals, including a member of the Royal family (the Prince of Wales) and media figures, who were personally identified with, or at least strongly sympathetic towards, the critical campaign
- competition in the press, which attracted several newspaper editors to the idea of campaigning on what they took to be a populist cause. The *Daily Mail* launched an explicit campaign under the editorial headline 'An issue of concern to every reader' on 6 February, while the *Independent on Sunday* began its campaign the following day with the front-page headline: 'Stop GM Foods: Modified Crops "Out of Control".'

(POST 2000a)

Evidently this last factor proved particularly consequential. Far from adopting a straight 'reporting' stance, the study maintains, all of the tabloid newspapers and several of their broadsheet counterparts developed a 'campaigning' approach to the issue. Some launched a formal campaign, others allowed a campaigning style to remain implicit, but taken together these newspapers 'set the agenda' for the most turbulent phase of the debate (two of the most 'strident' being the *Daily Mail* and the *Express*, locked in a circulation war at the time). Indeed, the study indicates that the broadcast media, especially BBC Radio 4's highly influential *Today Programme*, frequently took their cue from press headlines.

Significantly, the campaigning newspapers succeeded in pursuing the dispute far beyond the narrow limits ordinarily associated with a 'science/technology story'. While the overall amount of the coverage being generated by the campaigning newspapers was broadly equivalent with the non-campaigning titles, its form was markedly different. According to the study, the campaigning newspapers 'entered the debate first, raised new issues first, made use of more dramatic headlines, and devoted a larger proportion of their coverage to commentary (rather than news or features), especially in the early stages of the debate'. Similarly important is the finding that non-scientific (that is, general, political, environmental or consumer) correspondents contributed the majority of news items, with specialist science correspondents never being responsible for more than 15 per cent of the total news coverage at any one stage during the period under scrutiny. Indeed, during 11–12 February, the key point where the GM food story broke, none of the news articles in the 11 national newspapers surveyed had been written by science journalists (in contrast, 45 per cent had been written by political journalists). Science journalists were far more likely to be producing feature articles, particularly in the broadsheets. Hence one reason, the study suggests, why many members of the scientific and science policy communities regarded so much of the coverage to be 'unscientific' or even 'anti-scientific' in character.

Politicians were similarly quick to criticize the news coverage, more often than not insisting that it was sensationalist. Headlines cited by the study provide a flavour of the tone sometimes adopted:

> Alarm over 'Frankenstein Foods'
>
> (*Daily Telegraph*, 12 February 1999)
>
> THE PRIME MONSTER: Fury as Blair says: 'I eat Frankenstein food and it's safe'
>
> (*Mirror*, 16 February 1999)
>
> GM Foods: How Blair ignored our top scientists
>
> (*Daily Mail*, 18 February 1999)

> Mutant crops could kill you
>
> (*Daily Express*, 18 February 1999)
>
> Gene crops could spell extinction for birds
>
> (*The Guardian*, 19 February 1999)

Writing in the *Daily Telegraph*, Prime Minister Tony Blair complained of 'two weeks of misinformation', insisting that he and his family happily ate GM food (Greenpeace responded, dumping four tons of GM soya beans outside his Downing Street residence). One of his cabinet ministers, Jack Cunningham, accused the media of 'mass hysteria' as one supermarket chain after the next removed GM products from its shelves. Meanwhile, by May 1999 the Royal Society – representing Britain's scientific establishment – had declared that in its view 'no conclusions should be drawn' from Pusztai's experiments. His work, it alleged, was 'flawed in many aspects of design, execution and analysis'. This verdict, based on a review of the research by the Royal Society's own experts, went on to stress that while harmful effects from genetic modification could not be 'categorically ruled out', further research was required. In the meantime, the Society stated that it would not be endorsing the British Medical Association's (BMA) call for an indefinite ban on the commercial planting of GM crops until their effects were better understood. Particularly worrying, in the BMA's view, was the use of antibiotic-resistant genes in foods, the fear being that they may increase antibiotic resistance in humans. The BMA was also pressing for a ban on imported GM foods if they were not clearly labelled as such (*The Guardian*, 18 May 1999). Both the Royal Society and the BMA appeared to agree, however, that no evidence of harmful effects from genetic modification does not prove, in turn, that it is safe. Where they appeared to differ is over the relative assurances that it is likely to be no more dangerous than other forms of food.

The overall tenor of public opinion in the UK, the POST (2000a) study suggests, appeared to have shifted from cautious approval of GM food in 1996 to sceptical disapproval in late 1999 (see also Reiss and Straughan 1996; Durant et al. 1998; Hornig Priest 2001). Of crucial importance here, once again, was the influence of the campaigning newspapers. This type of journalism, the study contends, politicized the coverage of GM food such that it gave 'the debate its characteristically confrontational and even raucous qualities'. Still, it cautions, the opportunity for this campaign was provided by 'the steady divergence after 1996 between governmental and industrial policy on GM food, one the one hand, and public opinion on the other'. This divergence, then, was the key factor. Several additional factors followed from it:

- even a single, unpublished (and therefore unauthenticated) scientific claim – given the 'right' circumstances – can have an extraordinary impact on public debate and public opinion
- dealing with expert disagreement in socially sensitive areas of scientific research is extremely difficult
- when scientific or science-related issues become high profile news, events can move very quickly indeed – and not always in directions that scientists expect
- while low-profile science stories in the news are often handled by specialist science or technology correspondents, high-profile science stories are often the province of a wider range of journalists – up to and including newspaper editors. When this happens, the ways in which science is handled can also change significantly
- for the most part, science and science-related issues are the subject of *reporting* in the media. Occasionally, however, such issues can become the target of campaigning. The rules of engagement of science and scientists with the media are completely different under conditions of reporting and campaigning.

(POST 2000a)

How, then, to enhance the quality of news coverage about scientific issues so as to help enrich the larger public dialogue about their significance? In light of the POST study's findings, the House of Lords Science and Technology Committee concluded that those journalists who are not specialized science correspondents must undertake steps to improve their handling of the science aspects of news stories. This challenge, members of the committee believed, could be met via a set of suggested guidelines intended to 'change the behaviour of the media' while, at the same time, 'changing the behaviour of the scientists in dealing with the media'.

Turning first to the media, the report recommended that the Press Complaints Commission (PCC) adopt and promulgate the Royal Society's news guidelines for editors. Aimed at both specialist and non-specialist science correspondents, these suggestions cover the issues of 'accuracy, credibility, balance, uncertainty, legitimacy, advice and responsibility'. Additional mention is made of the issue of 'risk', with the report pointing out that particular care needs to be taken with the word 'safe' in this context. Given that some risks are acceptable, while others are not, the report contends: 'The very question "Is it safe?" is itself irresponsible, since it conveys the misleading impression that absolute safety is achievable. It also defeats its own purpose, since the only possible answer is "No". A better question is "How dangerous is it?" or "Is it safe enough?" ' (POST 2000a; see also Hargreaves and Ferguson 2000).

Regarding the guidelines for scientists, the report endorses a statement made by one of its witnesses: 'In a democratic media culture scientists have to learn to take the rough with the smooth like everybody else.' Far from implying a desire to maintain the status quo, however, the committee proceeded to express its conviction that 'the culture of UK science needs a sea-change, in favour of open and positive communication with the media.' Attention is drawn to the need for the necessary training and resources being made available to enable the scientific community to help non-specialists in the media cover scientific stories more satisfactorily. Clearly, much work remains to be done.

Lessons learned?

It is all too apparent that BSE/vCJD, as well as GM crops and food, will continue to feature in the news for years to come. Lessons will continue to be drawn from the mistakes made to date. In the case of the BSE crisis, as Powell and Leiss (1997: 3–4) suggest, the most important lesson is that 'there is a terrible risk in seeking to comfort the public with "no-risk" messages'. While it is still early to say for the GM debate, the main lesson identified by the POST (2000a) report above is that 'in a democracy, any significant interest – science included – ignores the public at its peril.' Even in recognizing lessons such as these ones, however, searching questions remain about how best to improve the reporting of food scares. Particularly pressing is the need to make the scientific language of risk understandable to members of the public while, at the same time, making their perceptions of risk inform the basis of the science involved. Ordinary or 'lay' people have a crucial part to play in scientific decisions, and their opinions have to be seen to count. 'Risks are not simply questions of abstract probabilities or theoretical reassurances,' as Horton (1999), editor of the medical journal *The Lancet* points out. 'What matters is what people believe about these risks and why they hold those beliefs.'

It follows, then, that public trust is a prize to be won through open debate, and not something that can be assumed as a right by experts. In countries such as Britain, as Collins and Pinch (1998: 124) contend, 'the official response to public health risks has traditionally been paternalistic reassurance. The government judges that the danger of panic usually outweighs any real risk to its citizens'. Typically the main task becomes one of allaying public fears, a response that has clearly informed how the food crises have been handled. As Collins and Pinch maintain with regard to BSE, not only does such an approach damage public health, it also harms science:

> Could the disease be passed between species – sheep to cows and vice versa? Could it be the cause of a new variant of human brain disease? Was it safe to eat hamburgers? The contradictory answers coming from the scientific community, the attempts to conceal the whole truth, the political debacle as British beef was banned from export markets, the well-publicized collapse of British beef farming as indecisive policy followed indecisive policy: none of these were good for science. And yet in many ways the fault lay not with science, which was simply unable to deliver instant answers to some very subtle questions, but with the false expectations placed upon it. This allowed others to pass the blame to science whenever it suited them.
>
> (Collins and Pinch 1998: 154)

This tendency to pass the blame to science, emergent at a number of pivotal points as the respective BSE and GM food crises have unfolded, is arguably indicative of a cultural climate where science is frequently seen to be politics by another means. In this way, as Jasanoff (1998: 355) argues, these kinds of crises raise fundamental questions about the very legitimacy of state institutions and their advisory bodies. This is an age, she writes, 'of complex technological risks that defy full scientific understanding and paternalistic, topic-down control', hence the need to ensure adequate access for diverse voices in public dialogues about science. The policy process needs to be revitalized, she points out, so as to 'acknowledge multiple reservoirs of knowledge and expertise, recognize divergent perspectives on risk, and [to find ways to act] responsibly without necessarily knowing the full implication's of one's action' (Jasanoff 1998: 355; see also McNeil 1987; Myers 1990; Adam 1998, 2000).

Conceptual interventions of this type open up for interrogation the social contingencies of expertise in advantageous ways. Where expertise appears to be most taken for granted, it is at its most ideological. Different stakeholders, using contrary criteria, will each strive to invoke their preferred definition of expertise as being self-evidently authoritative. Such claims to expertise, despite appealing to the proclaimed rationality of science, are necessarily socially situated, perspectival and therefore politicized. Indeed, as the BSE and GM food crises demonstrate, whether such a claim is sustained or not may have as much to do with political and economic interests as it does with the apparent facts of the matter. Important questions thus arise as to the means by which certain claims to expertise come to be aligned with specific definitions of risk at the level of 'the best available scientific advice', to use a phrase favoured by some politicians. The work of documenting this process of convergence, it follows, may contribute to the negotiation of strategic realignments of expertise and risk along far more responsible, publicly accountable lines.

Democracy requires a robust exchange of viewpoints, and a journalism up to the challenge of giving them vigorous expression. As this chapter has shown, new forms of dialogue about food risks need to be fostered, particularly where the absence of scientific certainty becomes controversial. Nevertheless, even though the costs of recent failures to attend to public perceptions of risk vis-à-vis BSE and GM foods are still mounting, the soliloquies of experts continue to set the normative limits demarcating the field of debate.

Further reading

Adam, B. (1998) *Timescapes of Modernity: The Environment and Invisible Hazards*. London: Routledge.

Baker, R. (2001) *Fragile Science*. London: Macmillan.

Beardsworth, A. and Keil, T. (1997) *Sociology on the Menu*. London: Routledge.

Durant, J., Bauer, M.W. and Gaskell, G. (eds) (1998) *Biotechnology in the Public Sphere*. London: Science Museum.

Ford, B.J. (2000) *The Future of Food*. London: Thames and Hudson.

Hornig Priest, S. (2001) *A Grain of Truth: The Media, the Public, and Biotechnology*. Lanham, MD: Rowman and Littlefield.

Powell, D. and Leiss, W. (1997) *Mad Cows and Mother's Milk: The Perils of Poor Risk Communication*. Montreal and Kingston: McGill-Queen's University Press.

Rhodes, R. (1998) *Deadly Feasts: Science and the Hunt for Answers in the CJD Crisis*. London: Touchstone.

8

FIGURES OF THE HUMAN: ROBOTS, ANDROIDS, CYBORGS AND CLONES

We should be on our guard not to overestimate science and scientific methods when it is a question of human problems.

(Albert Einstein)

We now have discrimination down to a science.

(Vincent Freeman, in the film *Gattaca*)

What does it mean to be human? This question is hardly a new one, of course. It has been provoking intense moral and ethical debates for centuries. Still, in posing it from the perspective of today, a host of intriguing resonances may be discerned that take familiar philosophical debates in new directions. Precisely what it means to be human, some commentators argue, has never been quite so worryingly uncertain. Scientific and technological innovations, they point out, have begun to blur the boundaries between human beings and the machines they have created, between the living and the inanimate. 'If you have been technologically modified in any significant way,' Gray (2001: 2) maintains, 'from an implanted pacemaker to a vaccination that reprogrammed your immune system, then you are definitely a cyborg.' Some commentators go much further, warning that biotechnological advances – from artificial hearts to genetic engineering and cloning – are radically reconfiguring what counts as 'human' in unexpected, sometimes alarming ways. Will it soon be possible, they ask, to download human consciousness into a computer? For them, the era of the post-human beckons, for better or – as many would have it – for worse.

This chapter takes up this question of what it means to be human from a number of different vantage points. First our attention turns to the tale of *Frankenstein* due to its status – as Turney (1998: 3) argues – as the 'governing myth of modern biology'. The salience of this myth, and the extent to which it is drawn upon as a rhetorical resource to both advance and challenge certain formulations of science, receives particular attention. We then examine the respective figures of the 'robot', 'android' and 'cyborg', so as to discern how familiar boundaries between 'human self' and 'mechanical others' have been recast from alternative perspectives within science fiction. Next, the *Brave New World* of 'test-tube babies', along with genetically engineered 'designer children', is centred for discussion. It appears that earlier anxieties about humans being taken over by robots (echoes of the Frankenstein myth) are being replaced with fears that humans will be genetically engineered into becoming robot-like themselves. The chapter is then brought to a close by considering the prospect of human cloning. Whether science is out of control (*Frankenstein*) or in control (*Brave New World*), the possible implications of cloning people – that is, creating babies that are genetic replicas of adults – could hardly be more profound for humankind.

Frankenstein's monster

Mary Shelley's *Frankenstein: or, The Modern Prometheus* is a chilling novel, its poignantly disturbing vision of how the obsessive pursuit of scientific discovery may engender tragic consequences is likely to leave a lasting impression upon today's reader. First published in 1818, it offers a range of fascinating insights into the 'new and almost unlimited powers' of early-nineteenth-century science (especially with regard to electricity and chemistry) in relation to the moral climate of the time. These insights have led some commentators to suggest that the novel represents the prototype of science fiction as a literary genre. Aldiss (1973: 26), for example, suggests that it should be regarded as the first genuine science fiction story, inside of which 'lie the seeds of all later diseased creation myths'. He points out that Shelley, who had begun writing the story before her nineteenth birthday, 'had imbibed the scientific ideas of [Erasmus] Darwin and [Percy] Shelley; had heard what they had to say about the future; and now set about applying her findings within the loose framework of a Gothic novel' (Aldiss 1973: 21). Others have insisted that it belongs firmly within the 'horror' genre (a claim that has been reinforced by various Hollywood treatments over the years). Interestingly, at the time of its publication, the story was frequently classified as a 'romance' by critics anxious to dismiss it as a highly improbable flight of fancy, albeit one which invited moral depravity in their view (see Schoene-Harwood 2000). In any case, however, there appears to be little doubt that *Frankenstein* has come to occupy a central place in current media discourses concerned with biotechnology, the reasons for which are deserving of scrutiny here.

Before discussing the significance of the novel for these debates, though, it is worth pausing to provide a brief sketch of how scientific themes emerge in its narrative. The following passage helps to set the tone:

> It was on a dreary night of November that I beheld the accomplishment of my toils. With an anxiety that almost amounted to agony, I collected the instruments of life around me, that I might infuse a spark of being into the lifeless thing that lay at my feet. It was already one in the morning; the rain pattered dismally against the panes, and my candle was nearly burnt out, when, by the glimmer of the half-extinguished light, I saw the dull yellow eye of the creature open; it breathed hard, and a convulsive motion agitated its limbs.
>
> (Shelley [1818] 1994: 69)

The words belong to the young Swiss scientist Dr Victor Frankenstein. Determined to understand the secret of life, he has devoted two years to the ghastly work of collecting and assembling human remains from dismembered corpses taken from graveyards and charnel houses. His task complete, he is at once exhilarated and repelled by 'the hideous phantasm of a man' stretched out before him. Once he has succeeded in imparting in his creation the 'vital spark' of life (precisely how this is achieved is left unclear), Frankenstein immediately shrinks back in horror at what he has done:

> How can I describe my emotions at this catastrophe, or how delineate the wretch whom with such infinite pains and care I had endeavoured to form? His limbs were in proportion, and I had selected his features as beautiful. Beautiful! Great God! His yellow skin scarcely covered the work of muscles and arteries beneath; his hair was a lustrous black, and flowing; his teeth of pearly whiteness; but these luxuriances only formed a more horrid contrast with his watery eyes, that seemed almost of the same colour as the dun-white sockets in which they were set, his shrivelled complexion and straight black lips.
>
> (Shelley [1818] 1994: 69)

Despite this extraordinary achievement in 'infusing life into an inanimate body', Frankenstein is overcome by fear and revulsion. He flees to the countryside, hoping to forget all about the creation (upon whom he even fails to bestow a name). The now abandoned living being, denied the care and attention of his maker, is left to make his own way. Tragedy promptly ensues, as the creation is persecuted by virtually everyone he encounters due to his frightening appearance, leaving him utterly rejected by human society.

Eventually encountering Frankenstein himself, the creature implores him to construct a female companion for him to share his life. Frankenstein agrees to initiate the project, but soon reneges on the arrangement following the sudden realization that the two creatures, once united, might propagate between them a 'race of devils'. The creature is devastated by Frankenstein's decision to destroy his promised companion, vowing that 'I shall be with you on your wedding night!' before disappearing. The heart-broken creature, wretched with grief, subsequently makes good his threat, murdering his creator's fiancée. Frankenstein pursues his 'cursed and hellish monster' to the far north, believing that his creation should 'drink deep of agony' so that he may 'feel the despair that now torments me' (Shelley [1818] 1994: 256). Frankenstein fails to achieve his revenge, eventually succumbing to the cold. His final words, spoken to his friend Robert Walton, who has rescued him from the ice flow, are revealing. Bidding him farewell, Frankenstein declares: 'Seek happiness in tranquillity and avoid ambition, even if it be only the apparently innocent one of distinguishing yourself in science and discoveries. Yet why do I say this? I have myself been blasted in these hopes, yet another may succeed' (Shelley [1818] 1994: 275). The creature soon reappears at Frankenstein's bedside, full of remorse and recrimination. Confronted in turn by an enraged yet curious Walton, the creature admonishes him: 'My heart was fashioned to be susceptible of love and sympathy, and when wrenched by misery to vice and hatred, it did not endure the violence of the change without torture such as you cannot even imagine' ([1818] 1994: 278). He then declares his intention to collect his own funeral pile and 'consume to ashes this miserable frame, that its remains may afford no light to any curious and unhallowed wretch who would create such another as I have been' ([1818] 1994: 281).

In sharp contrast with several Hollywood filmmakers' depictions of this story, then, Frankenstein's creature is not inherently evil. Rather, it is the cruel experience of rejection that the creature suffers, especially with respect to his maker's failure to honour his moral duty to him, which motivates his violent fury. The novel, in my reading, is much less a critique of science than it is an indictment of scientists who refuse to accept responsibility for their actions. There are few passages which dwell on the actual science involved, and those which do offer little more than a broad affirmation of different scientific theories of the day, albeit with an imaginative twist. The plausibility of the novel's account of Frankenstein's creation of a living being in the laboratory rests on what Shelley describes as the application of certain imagined scientific facts and principles, as opposed to being based on an appeal to a larger supernatural power. It is this distinction between science and the supernatural, as several commentators have pointed out, that marks the novel's decisive break from earlier types of literary fantasy. Moreover, in the eyes of critics such as Aldiss (1973), herein lies the durability of the

Frankenstein legend. The novel, in his view, 'not only foreshadows many of our anxieties about the two-faced triumphs of scientific progress, it is the first novel to be powered by evolution'. That is to say, he adds, 'God – however often called upon – is an absentee landlord, and his lodgers scheme to take over the premises' (Aldiss 1973: 26; see also Gould 1996; Nottingham 2000).

Intrigued by the influence of Shelley's novel on popular images of biological science over the years since its publication, Turney (1998) pursues this line of inquiry in depth. *Frankenstein*, he argues, has become one of the most important myths of modernity, one with a life of its own. The concerns it identifies have proven to be central to longstanding debates. More than that, however, Turney maintains that the novel itself has become a versatile 'frame' or 'script' by which members of the public interpret their social relationships with science and technology. 'To activate it,' he writes, 'all you need is the word: *Frankenstein*' (Turney 1998: 6). The mere mention of the word is likely to evoke an entire story, thereby immediately recontextualizing any discussion in a new, sometimes startling way. It follows that the public's ready acceptance of the popular cultural script derived from the novel, retold in a myriad of ways in different media contexts over the years, is largely due to the fact that it 'expresses and reinforces an undercurrent of feelings about science' (1998: 36). In each rendering of the story a creation myth based on science is retained, but it is a science that is out of control to the point that it turns on the scientist. And yet, Turney observes, the myth rarely engenders a straightforward anti-science story. Rather, Frankenstein the scientist is typically portrayed as having understandable motives, even somewhat admirable ones, and audiences are invited to sympathize with both him and his creation.

For Turney, the *Frankenstein* script, 'in its most salient forms, incorporates an ambivalence about science, method and motive, which is never resolved' (1998: 35). This ambivalence is arguably nowhere more pronounced than in the realm of biological science, particularly where people's 'deepest fears and desires' about the 'violation of the body' reside. 'The human body', he writes, 'is both a stable ground for experience in a time of unprecedentedly rapid change and a fragile, limited vessel which we yearn to remake' (1998: 8). Shelley's imaginative leap has not only helped to fashion current images of science working to transform the body, but also given voice to 'fears of bodily intrusion as an implicit goal of the project of modernity' (Turney 1998: 218). To clarify this argument, Turney develops his notion of the *Frankenstein* myth as a rhetorical resource, showing how it is routinely exploited by proponents of different visions of the possible human benefits associated with biological science, as well as by their critics. Whether the issue under scrutiny is experiments on laboratory-fertilized human **embryos**, for example, or the long-term effects on people of genetically modified food ('Frankenfoods' in the eyes of some), the myth is recurrently drawn upon as a useful resource by diverse voices engaged in vigorous debate.

Not surprisingly, however, such discursive strategies are part of the problem, not the solution, in Turney's view. That is to say, he is convinced that they make the potential reconciliation of differing perspectives on how to control new technologies less likely to take place. The *Frankenstein* myth, he writes,

> invites an all-or-nothing response to a whole complex of developments, when we should be insisting on our right to choose some, and block others. When we do so, it should be for reasons which we can articulate more clearly than saying either that there are some things humans are not meant to know, or that we should not tamper with nature. But to prepare to articulate those reasons, first we must come to terms with *Frankenstein*.
>
> (Turney 1998: 11)

How best to recast debates otherwise framed around Frankenstein, with the 'yes' or 'no' questions and answers they tend to elicit, is a difficult challenge. The myth tends to polarize these debates in a manner which obscures their attendant complexities, offering instead a simplified conflict between dread and desire. New stories are required, Turney argues, so that people will be better placed to make informed, ethical decisions about techniques and possible applications. And yet, he concedes, the myth shows little sign of disappearing, as it is far too deeply embedded in cultural forms and practices. Indeed, not only will it continue to play a critical role in shaping public attitudes to science for a long time to come, he believes, but also as biological science develops there is every chance that its grip may even strengthen.

Further aspects of the Frankenstein myth emerge in the discussion of biotechnology below, but next our attention turns to the mechanical human as an object of the scientific imagination. Here, too, popular anxieties about the boundary between the human and the non-human is shown to have found formative expression in images first articulated in science fiction.

Of flesh and steel

The term 'robot' appears to have first surfaced in 1920 in the play *R.U.R.* (*Rossum's Universal Robots*) by the Czeck writer Karel Capek. Evidently 'robot' was derived from 'robota', loosely meaning 'drudgery' or 'servitude', and 'robotnik', a peasant or serf. The play premiered in Prague in 1921, and went on to enjoy worldwide success, including in London and on Broadway in New York (the English language translation having made some changes to the plot). Very briefly, the robots are the product of a mad inventor and his industrialist son living on an unidentified island, and are intended to serve as a source of cheap labour so as to replace human workers. Every once and awhile, however, a robot malfunctions. Most of the characters attribute the problem to a product defect, yet one character, Helena, sees in it evidence of an emerging soul. She works to help the robots, asking a scientist to effect modifications to them so that their souls might be more fully developed. Things quickly spiral out of control, with one of the modified robots issuing a manifesto: 'Robots of the world, you are ordered to exterminate the human race . . . Work must not cease!' This order is virtually completed – only one human, a scientist, escapes – before the robots realize that without 'the secret of life' they cannot produce other robots. Two chastened robots fall in love, are renamed Adam and Eve, and set out anew determined to avoid committing the sins that destroyed their predecessors.

The novel idea of the robot proved to be enormously popular with the play's audiences. Different accounts suggest, however, that Capek believed too much attention had been focused on the devices, at the expense of the larger social issues he was attempting to address. The substitution of robots for humans had allowed him to express allegorically, as Disch (1998: 9) observes, 'the moral truth that the industrial system treated human labourers as though they were machines, sowing thereby the seeds of an inevitable and just rebellion'. Audiences appeared to be much more interested in the idea that robots could be more perfect than humans, especially with regard to their superior intellects, and yet still succumb to emotions and feel pain. Here it is important to note that the robots in Capek's play were humanlike in appearance and organic in origin (as such, they would typically be described as androids today). Presumably this portrayal of robots was still fairly fresh in the public mind when, in 1926, Fritz Lang's silent film *Metropolis* was released. The film portrayed a mechanized society, and shared with Capek's play a pessimistic appraisal of technology as a dehumanizing force. Regarded by many commentators to be the first science fiction film, *Metropolis* borrowed certain elements from *R.U.R.*, not least the idea of a mad inventor manufacturing (out of metal in this case) a female 'robotrix' to be used to incite the workers to revolt. Its vision of the future is dystopian, despite its relatively upbeat ending, and yet regarded by many to be prophetic.

Russian-born US scientist and writer Isaac Asimov is generally credited with introducing the world to the word 'robotics', namely as a way to describe the technology of robots, in 1942. The term appears in his short story 'Runaround', published in the pulp magazine *Astounding Science Fiction* in March of that year. Asimov's interest in robots had been addressed in earlier stories, and would be developed further in later ones – eventually culminating in a widely acclaimed book-length collection, titled *I, Robot*, in 1950. It is in this series of stories that the idea of a 'positronic brain' appears as well. A positron, the unstable antiparticle of an electron, was envisaged by Asimov to be suitable material for the construction of an artificial brain. Partly in response to what he regarded as the technophobic 'Frankenstein syndrome' characteristic of so many science fiction portrayals of robots, he set down his three 'Laws of Robotics' (the 'zeroth' law being added later) to guide the brain's use of logic. The proposed laws, apparently formulated in discussions with his editor John W. Campbell Jr, were as follows:

- *Law Zero*: a robot may not injure humanity, or, through inaction, allow humanity to come to harm.
- *Law One*: a robot may not injure a human being, or, through inaction, allow a human being to come to harm.
- *Law Two*: a robot must obey orders given it by human beings, except where such orders would conflict with the First Law.
- *Law Three*: a robot must protect its own existence as long as such protection does not conflict with the First or Second Law.

Taken together, these laws form the basis of an ethical system, and as such should be fundamental to robotic design. As Clute and Nicholls (1999: 56) note, they 'helped put paid to the increasingly worn-out pulp-magazine convention that the robot was an inimical metal monster; [and] they allowed [Asimov] to create a plausible alternative for the 1940s in his positronic robots'. At the same time, the laws in 'lawyerly fashion' also 'generated a larger number of stories which probed and exploited various loopholes' (Clute and Nicholls 1999: 56). To the extent that other writers were willing to portray robots as having an in-built ethical system, Asimov believed, more subtly nuanced stories would emerge at a time when public attitudes to technology were becoming increasingly ambivalent.

Public disillusionment with technology grew ever stronger in the wake of the US military's atomic bombings of Hiroshima and Nagasaki (see Boyer 1985). Robot technology was no exception, and even Asimov himself was moved to write his first sinister-robot story, 'Little Lost Robot', in 1947. Still, he remained pro-robot in principle, a stance which did find support among several other writers. Female robots continued to prove particularly salient across the media, an exemplar of sorts being the one appearing in the 1949 film *The Perfect Woman*. Here a professor sets out to create a robot (or android in today's language) in the image of his niece. 'I call her the perfect woman,' the inventor declares, 'she does exactly what she's told, she can't talk, she can't eat, and you can leave her switched off under a dust sheet for weeks at a time.' The film's sexism, as Chibnall (1999: 58) remarks, reduces Olga, the robot, to the status of a decorative toy for men, thereby giving expression to male domination fantasies. In the end, Olga malfunctions at the mention of the word 'love', issues forth sparks and then smoke, and promptly explodes (see also Wolmark 1999).

During the 1950s, the robot quickly became a popular mainstay of science fiction. Indeed the proliferation of robot-centred storylines has been such that to even sketch the broad features of its history would be a daunting task. Any such history would discuss the influence of the robot called Gort in the 1951 film *The Day the Earth Stood Still*, as well as Robby the Robot, who was enormously popular in films such as *Forbidden Planet* (1956) and *The Invisible Boy* (1957). Robby, as O'Brien (2000: 29) observes, was 'one of the great icons of the science fiction genre'. The next decade would see the Daleks from *Doctor Who* appear (although they had a humanoid brain inside their metallic bodies), as well as the robot in the television series *Lost in Space* whose catch-phrase was 'That does not compute'. Similarly significant were the robots in *Silent Running* (1971), *Sleeper* (1973), *Star Wars* (1977), where R2D2 features (as does the android C3PO), and its sequels, *Saturn 3* (1980), and many others since. The current BBC series *Robot Wars* would also deserve mention. Speaking in general terms, though, the robot has been typically portrayed as a dangerous, menacing machine that eventually turns on its maker (as noted above, the Frankenstein myth is an enduring one).

'I propose that we build a robot who can love . . . a robot that dreams.' Such is the challenging vision set down by a scientist in the 2001 film *A.I.: Artificial Intelligence*. He and his colleagues at Cybertronics Manufacturing promptly begin work and, two years later, complete an **AI** prototype of a so-called perfect child, 'always loving, never ill, never changing'. A human couple adopts the robot child, who they name David, and welcome him into their family as a replacement of sorts for their son Martin, who is in a coma. To their surprise, however, Martin suddenly recovers, and is resentful of his new 'mecha' (mechanical) brother. Eventually, as the relationship sours due to a variety of mishaps, David is abandoned in a forest, where he struggles to survive with the help of similarly forsaken robots. Several harrowing scenes ensue, particularly those revolving around the 'flesh fair' where robots are publicly sacrificed in a medieval-like circus of torture. 'They hate us, you know' one robot tells David as together they witness fellow robots being scalded with acid, quartered, pulverized or turned into a 'robot cannonball' to the delight of the cheering 'orgas' (organic beings or humans) who detest them. Fearful that mechas will take over the world, humans have become almost as coldly mechanical and unfeeling as the technology upon which they now depend. *A.I.*, a joint venture between the late Stanley Kubrick and Steven Spielberg, was based on Brian Aldiss's short story, 'Supertoys Last All Summer Long'. It intermingles science fiction with fairy tale (David, like Pinocchio, longs to find the Blue Fairy who, he believes, will transform him into a real boy) in what becomes a tragic vision of what becomes a robot equipped with real feelings, such as love. Even if a robot could learn to love humans, the film asks, could humans learn to love a robot?

Although the term robot is used in this film, contemporary writers of science fiction would be more likely to use the term 'android' as David is an artificial human. Having said that, however, for many writers the two terms are interchangeable, particularly where the figure in question is composed of both synthetic and organic materials. For example, the figure of Data in *Star Trek: The Next Generation*, as noted in Chapter 2, is similarly struggling to become fully human, thereby effectively blurring the division between machine and sentient being. Viewers of the programme are informed in the episode 'Datalore' that he was initially manufactured in an attempt to bring 'Asimov's dream of a positronic robot' to life (evidently the series' producer had asked for Asimov's permission to borrow the idea). Data shares with an Asimovian robot an in-built ethical system, but does not always respect the 'Laws' (noted above) where necessity dictates an alternative course of action. Interestingly, one episode revolves around his discovery that his creator, Dr Noonien Soong, had included an 'emotion chip' in his electronic circuitry. While the chip can be 'turned off' when necessary, its presence promises to bring him one step closer to realizing his desire to become human. This same desire is exploited on several occasions, although perhaps most chillingly by the Borg Queen in the 1996 film *First Contact*. Having reactivated Data's emotion chip, she proceeds to graph organic skin onto his arm, promising to cover his entire body if he agrees to surrender the *Enterprise*'s encryption codes (the codes would provide her with control over the starship's main computer). 'You are an imperfect being, created by an imperfect being,' she tells him. 'Finding your weakness is only a matter of time.' While it appears that Data may be willing to betray his *Enterprise* colleagues and join the Borg, it soon becomes clear that he is simply waiting for the right moment to overpower his enemy. Once this has been achieved, he duly informs his Captain that indeed he had been tempted by the Queen's offer. For how long? Data declares: '0.68 seconds, sir. For an android, that is nearly an eternity.'

Fans and other followers of the various *Star Trek* series continue to debate how best to define Data's ontological status (similar debates transpire over the status of 'replicants' in the film *Blade Runner*; see Sammon 1996). For some he is simply a robot, for others an android, and for still others a cyborg. With regard to the latter, the case for Data's cyborgization rests largely on the fact that his consciousness is primarily made up of memories borrowed from humans living near the laboratory where he was assembled. A cyborg, as Gray (2001) suggests, is a self-regulating organism that combines the natural (evolved, living) and the artificial (made, inanimate) together into one system. Almost all of us, by this definition, have been 'cyborged' to some extent – indeed some researchers go so far as to suggest that simply by wearing glasses one qualifies for cyborg status. In any case, credit for first coining the term 'cyborg' is usually given to Manfred Clynes. In 1960 he was working on ideas about humans in space at NASA when, according to Gray, he contracted 'cybernetic' and 'organism' into 'cyborg' in order to help enliven a research paper he was preparing with psychologist Nathan Kline. Their paper's rather bold thesis was that humans might be effectively modified with implants and drugs so as to allow them to exist in space without the benefit of space suits. From there the term 'cyborg' began to catch on, albeit not within the scientific community (who, Gray writes, 'preferred more specific labels such as biotelemetry, human augmentation, human-machine systems, human-machine interfaces, teleoperators, and – to describe copying natural systems to create artificial ones – bionics': Gray 2001: 19). Instead it was left to others to popularize the term, such as David Rorvik with his 1971 book *As Man Becomes Machine*. Much is made there of how human and machine are 'melding' into a 'new era of participant evolution'. Incidentally Rorvik, a US science writer, would later spark considerable public controversy by claiming to have witnessed the cloning of a human being (see Featherstone and Burrows 1995; Pence 1998, Van Dijck 1998).

Science fiction writers, long before the first 'cyberpunk' novel was written, took up and promptly elaborated upon the term 'cyborg', and in so doing ensured its popular currency. By and large, many of them saw in it a useful way to pinpoint how technology was being integrated into natural systems – as opposed to the reverse of this process, more typically associated with androids. The ideas behind the term had long been explored in a variety of science fiction texts, with E.V. Odle's (1923) *The Clockwork Man* widely regarded as the earliest major cyborg novel. 'Pulp' magazine writers similarly explored related themes, with forerunners including the *Captain Future* series (see Ashley 2000). More recent variations on this theme include Martin Caidin's 1972 novel *Cyborg*, which was adapted into television's

The Six Million Dollar Man in 1973 (followed by *The Bionic Woman* in 1976). Cinematic treatments include *The Terminator* films (1984, 1991), where Arnold Schwarzenegger plays a time-travelling cyborg, as well as *Robocop* (1987) and its sequels (1990, 1993), *Cherry 2000* (1987), the *Cyborg* films (1989, 1993, 1995), *Total Recall* (1990), *Eve of Destruction* (1991), *Universal Soldier* (1992) and *Inspector Gadget* (1999) among several others. To varying degrees, each film engages with the contradictory tensions engendered via the clash of humanization and mechanization processes as boundaries are crossed, criss-crossed and crossed again.

The fluidly contingent dynamics of power are typically pivotal to cyborg narratives. Technological imperatives are recurrently shown to recast familiar notions of human agency and control, often in ways as disturbingly nihilistic as they are imaginatively spectacular. Cyborg imagery, as Haraway (1991) argues in her well-known 'Cyborg Manifesto' essay, is important because it 'can suggest a way out of the maze of dualisms in which we have explained our bodies and our tools to ourselves' (1991: 181). That is to say, she believes that this kind of imagery has the potential to encourage people to challenge the traditional Western dualisms upon which certain logics and practices of domination rely. She is thinking here of a range of dualisms, including: 'self/other, mind/body, culture/nature, male/female, civilized/primitive, reality/appearance, whole/part, agent/resource, maker/made, active/passive, right/wrong, truth/illusion, total/partial, God/man' (Haraway 1991: 177). Such an approach, it follows, is sensitive to the heteroglossic array of voices – especially where issues of gender, class, ethnicity, sexuality and so forth are concerned – so frequently silenced by so-called universal, totalizing theories. A cyborg politics thus refuses to uphold the seemingly neat and tidy distinction between a human 'us' and the machine as 'other', in part by bringing to the fore questions of experience and embodiment otherwise left unrecognized. The machine, in Haraway's view, 'is not an *it* to be animated, worshipped, and dominated'. Rather, she writes, the 'machine is us, our processes, an aspect of our embodiment. We can be responsible for machines, *they* do not dominate or threaten us. We are responsible for boundaries, we are they' (1991: 180).

Test-tube babies, designer children

'And this,' said the director opening the door, 'is the Fertilizing Room.' With a welcoming sweep of his hand, he invited his guests to admire the laboratory before them. Their eyes took in the incubators, microscopes, and racks upon racks of numbered test tubes, the latter containing the week's supply of ova (being held at 'blood heat') and male gametes. Describing for them 'the liquor in which the detached and ripened eggs were kept', the director proceeded to show them how:

> this liquor was drawn off from the test-tubes; how it was let out drop by drop on to the specially warmed slides of the microscopes; how the eggs which it contained were inspected for abnormalities, counted and transferred to a porous receptacle; how (and he took them to watch the operation) this receptacle was immersed in a warm bouillon containing free-swimming spermatozoa – at a minimum concentration of one hundred thousand per cubic centimetre, he insisted; and how [several steps later] the fertilized ova went back to the incubators; where the Alphas and Betas remained until definitely bottled; while the Gammas, Deltas, and Epsilons were brought out again, after only thirty-six hours, to undergo Bokanovsky's Process.
>
> (Huxley [1932] 1955: 16–17)

Precisely what Bokanovsky's Process entailed, he promptly explained:

> One egg, one embryo, one adult – normality. But a bokanovskified egg will bud, will proliferate, will divide. From eight to ninety-six buds, and every bud will grow into a perfectly formed embryo, and every embryo into a full-sized adult. Making ninety-six human beings grow where only one grew before. Progress.
>
> (Huxley [1932] 1955: 17)

Progress, indeed. This memorable scene appears in Aldous Huxley's *Brave New World*, a novel written in 1931 about events he imagined taking place over six hundred years into the future. It envisages a world where a brutal political state controls human reproduction, using foetal hatcheries to breed children in accordance with society's specific requirements. Alphas will occupy the peak position in the predetermined intellectual class system, for example, while epsilons will be relegated to the lowest rank. Aligned with each position are particular social roles to be occupied when the 'uniform batches' of children have become 'standard men and women'. The very power to determine human reproduction has become 'one of the major instruments of social stability'. Propaganda, mood-altering drugs ('soma'), multisense movies ('feelies') and officially encouraged sexual promiscuity similarly help to ensure that everyone knows their proper place, and acts accordingly.

As *Brave New World* unfolds, it soon becomes apparent that the reproductive technologies being described anticipate the development of human cloning (the topic of the next section of this chapter). Huxley provides few additional details to clarify the actual processes involved; indeed, what references there are make the procedures sound more closely aligned with the shopfloor than with genetic engineering. 'The principle of mass production', the director declares, 'at last applied to biology.' What is in effect an assembly line – the 'hum and rattle of machinery faintly stirred the air' – is shown to operate with unfailing precision as the eggs undergo treatment, each set to develop into another dehumanized cog in a nightmareish society's wheel. Interestingly, writing in a foreword for a 1946 edition of his novel, Huxley observed that in the story 'this standardization of the human product has been pushed to fantastic, though not perhaps impossible, extremes. Technically and ideologically we are still a long way from bottled babies' ([1932] 1955: 13). As it turned out, of course, technological feasibility merged with ideological justification much sooner than Huxley might have dreamt possible.

The birth of the world's first 'test-tube baby' took place just over three decades later. On 25 July 1978, Louise Brown was born to proud parents Lesley and John Brown, in Oldham general hospital in Lancashire, England. Baby Louise was delivered by caesarean section, and weighed in at 5 pounds, 12 ounces. Her birth had been made possible by years of experimentation conducted by research scientist Robert Edwards and gynaecologist Patrick Steptoe. Together they had removed an egg from her mother's ovary, fertilized it in a test tube with sperm from her father, and then implanted back into her mother. The result, in their words, was the birth of a 'normal healthy infant girl'. The announcement of the remarkable success of this 'test tube' procedure, or more accurately *in vitro* fertilization (IVF) in a laboratory Petri dish (*in vitro* is Latin for 'in glass'), astonished the world. The phrase 'test-tube baby', with all of its *Brave New World* connotations, instantly became the preferred phrase in the public lexicon. Journalists had been arriving from around the globe to cover the breakthrough, desperately trying to help their audiences comprehend the seemingly incomprehensible. Some of them had attempted to bribe hospital staff for information, others used a local detective agency, while others sought to pass themselves off as plumbers or window cleaners to gain entry into the building. The police were called to clear news photographers away from the windows. The journalistic siege of the hospital, as might be expected, promptly became a news event in itself (see Karpf 1988; McNeil et al. 1990; Van Dijck 1995; Barnard 1999).

News coverage of the 'miracle baby' or the 'baby of the century', as some journalists called her, remained extensive for days afterward. One newspaper, the *Daily Mail*, paid the Browns for exclusive rights to the story (reputedly £300,000), thereby intensifying the frantic struggle of each journalist to develop a unique angle. The ensuing public debate, the rudimentary features of which had been assuming shape in the weeks prior to the announcement, became increasingly charged as different voices weighed in. Differing perceptions of the likely implications of IVF generated considerable controversy, much of it expressed in moral, ethical and religious terms. Huxley's novel, as might have been anticipated, provided something of a background 'script' for contending premonitions of the future. An item in the *Daily Mail*, for example, states:

> even as we share the happiness of John and Lesley Brown . . . we cannot fail to be aware of the unease some genuinely feel.
>
> Amid the rejoicing there are those who shiver involuntarily. 'Where,' they ask, 'is it going to end?' And implied in that question is a chill premonition of a brave new world coming closer.
>
> (*Daily Mail*, 27 July 1978; cited in Turney 1998: 183)

Similarly, a columnist in a rival tabloid newspaper, the *Daily Express*, saw it as his duty to alert readers to the possibility of 'genetic engineering on a massive scale performed by scientists working under the direction of the state'. This process, he warned, might already be underway:

> It could very easily happen . . . we already have sperm banks and may soon have egg banks. Under the guise of eliminating hereditary diseases and defects, governments could easily insist on the 'right' or 'best' kind of genetic material to be used . . . Far fetched? Of course it is far fetched. But the birth of the Brown baby in Oldham has already fetched us far along that road.
>
> (*Daily Express*, 27 July 1978; cited in Turney 1998: 184)

His conclusion was repeated in enlarged type, centre page: 'If I peer into the future I see a society of people bred to measure.' According to Turney's (1998) study, doubts and suspicions were aired across the British and international media, and frequently given shape by interpretative frames that owed much to the novel. Here he cites as evidence a line from *Newsweek* magazine's coverage: 'She was born at 11.47 p.m. with a lusty yell, and it was a cry heard round the brave new world.' Meanwhile in *Time* magazine's cover story: 'To millions of people in Britain and elsewhere around the world last week, it seemed as if Huxley's prophetic vision had become reality' (*Time*, 31 July 1978).

For some critics of IVF procedures, conception outside of a woman's body was an unnatural, even irreligious, act that should be avoided at all costs. Some of its advocates, in contrast, refused to see what the fuss was all about, insisting instead that it was simply a new way to help infertile couples have a baby, nothing more. It would be the latter definition which would prevail. Despite the misgivings of critics, the procedure quickly became normalized in the years to follow as fertility clinics began to appear in remarkable numbers. Still, the shadow cast by science fiction depictions of scientific conduct, especially *Frankenstein* and *Brave New World* but also George Orwell's *Nineteen Eighty-Four*, refused to disappear entirely from the media coverage. For example, almost ten years later, as the British government moved to legislate regulations for research involving human IVF embryos, some journalists did not hesitate to invoke Shelley's imagery. *The Sun* newspaper used a still from the film *Frankenstein* to accompany its news account, while *Today*'s item began as follows:

> CLAMP ON FRANKENSTEIN SCIENTISTS
>
> SCIENTISTS are to be banned by law from creating superbeings in the laboratory.
>
> The technique, known as cloning, which can produce identical humans from a single cell, will also be made a criminal offence.
>
> The Government admits that the prospect of Frankenstein-style experiments is unlikely, but it wants to stop any genetic tinkering with embryos which would pre-determine characteristics.
>
> A White Paper published yesterday proposes the clampdown on test-tube baby experiments.
>
> (*Today*, 27 November 1987; cited in Mulkay 1997: 119)

Mulkay's (1997: 119) reading of the news coverage at the time leads him to argue, in turn, that 'Frankenstein's prominence suggested strongly to readers that, despite official disclaimers, these scientists were dangerous and must be held on a tight rein'. Since then, these kinds of fears around IVF techniques have largely subsided, although they have continued to resurface on occasion with regard to issues such as 'surrogate mothers' or, more recently, the media storm over a British woman's efforts to use her dead husband's sperm to conceive (see Hartouni 1997; Harris 1998). Perhaps most controversial of all, however, is the prospect of genetic scientists acquiring the power to 'design' babies 'to order'. While in many countries it is currently legal for human embryos to be scanned for various genetic disorders, advances in IVF suggest that soon scientists will be able to add characteristics to a newly fertilized embryo so as to produce certain 'desirable' characteristics while eliminating 'undesirable' ones.

'There is no gene for the human spirit' is the marketing catch-phrase for the film *Gattaca* (1997), to my mind one of the most interesting depictions of how genetic technology might one day be routinely applied to humans.

Set in the 'not too distant future', the society it portrays is ruled by 'designer people', their genes having been refashioned in test tubes (incidentally, the word 'Gattaca' itself is composed of the letters used to label the nucleotide bases of DNA). Control over genetic material is primarily a question of economic power, with wealthier people able to afford access to its proclaimed benefits (physical attractiveness, athletic prowess, a genius intellect) for their children in a way denied to the financially disadvantaged. Throughout the film, the different characters draw upon a particular lexicon, including words such as:

- *Valid*: an individual who is a member of the genetic elite; he or she has a high 'grade' or 'G.Q.' (Genetic Quotient). Slang for Valid is 'Vitro', a 'Made Man' or a 'Dan' – the acronym DNA having been altered to 'Dan'.
- *In-valid*: an individual who is a member of the 'genetic underclass'; he or she has had a 'natural birth'. The word is pronounced as 'invalid'. Slang equivalents include: 'genojunk', 'uttero' and 'blackjack birth'.
- *De-gene-erate*: an In-valid who assumes the genetic identity of a Valid, usually with the assistance of a 'DNA Broker', in order to avoid detection and discrimination. Also called a 'borrowed ladder'.
- *Genoism*: the official term for genetic discrimination. While technically illegal, an individual's position in the social hierarchy is identified once their 'genetic quotient' is determined at birth.
- *Hoover*: slang for members of Gattaca's law enforcement personnel, who typically vacuum up organic matter as evidence at crime scenes. Synonyms include 'cleaners' or 'J. Edgars' (the latter a play on J. Edgar Hoover, onetime leader of the FBI in the US).

The film's plot revolves around an outsider named Vincent Freeman who, having had a so-called natural birth, is duly categorized as an 'In-valid'. In a flashback scene to the precise moment of his birth, a nurse is shown taking a blood sample from his heel. The results: 'Neurological condition 60 percent probability, manic depression 42 percent probability, attention deficiency disorder 89 percent probability . . . Life expectancy 30.2 years.' The film moves forward in time to the scene where Vincent's parents, having decided to have a second child, discuss its possible genetic make-up with a friendly geneticist. First they decide on its sex, then specify eye colour, hair type, skin colour, and so forth. The geneticist promises to ensure that the child will have 'no critical predisposition to any major hereditary diseases'. Even more, he will take the necessary measures to 'eradicate any potential prejudicial dispositions like premature balding, myopia, domestic violence, obesity, alcoholism and addiction susceptibility'. Vincent's parents, uneasy with such decisions, are offered reassurance by the (black, balding) geneticist: 'We want to give the child the best possible stuff, believe me, there is enough imperfection built in already. Think of the child as yours, simply the best of you . . . You could conceive naturally a thousand times and never get such results.'

In a world characterized by the prejudicial imperatives of genetic elitism or 'genoism', then, obtaining the ideal of human perfection is a matter of engineering. Vincent works at the Gattaca Aerospace Corporation as a cleaner, describing his place in society (via a voice-over) by stating: 'I belong to a new underclass, no longer determined by social status or the colour of your skin. We now have discrimination down to a science.' Far from accepting his 'genetic destiny', however, Vincent yearns to fulfil his childhood dream of being an astronaut. To achieve this end, he conspires with Jerome Morrow – a 'Valid' – to forge a new identity for himself, namely by using samples of Jerome's blood, urine, skin and hair to escape detection as an In-valid. Entry into the Gattaca building each day requires Vincent to pass a blood test, for example, so he wears a false finger tip containing a sample of Jerome's blood. He also hides on his person a pouch of Jerome's urine, tiny flakes of his skin, strands of hair, nail clippings, and so forth, so as to help sustain the genetic illusion or, in Vincent's words: 'to limit how much of my In-valid self I would leave in the Valid world.' As the realization of his dream to travel in space edges closer, Vincent must cope with a police investigation for an unsolved murder that threatens to reveal his concealed identity. He also meets and falls in love with Irene, a Valid but one who is classified at Gattaca as having 'minor shortcomings' in genetic terms. Unknown to him, she has taken a hair from a comb in his desk drawer (the hair is actually one of Jerome's) and has had it 'sequenced'. The testing complete, the computer readout confirms for her that he will make a genetically acceptable, even desirable mate. Thus what for Vincent is a romantic attraction is for Irene, at least initially, a process of partner choosing undertaken with the precision of 'science'. As the film proceeds to demonstrate, however, 'there is no gene for fate'.

Over the course of the last few years since the film's release, several of the insights it provides into the possible human consequences of genetic engineering have become even more pertinent. Looking across the media field, there has been a marked rise in the number of references to genes and DNA appearing as the language of biotechnology slowly permeates public culture. Recent examples include news reports that a genetic basis has been discovered for a given character trait – such as sexual orientation, intelligence, aggression, language acquisition or a predisposition to happiness – as well as various physical characteristics, such as hair and eye colour, or obesity. In many such reports, voices are proclaiming that soon it will be possible to suitably 'modify' genes as required, thereby improving people's life chances. Not surprisingly, however, other scientists have called into question the evidential basis and validity of research into behaviour genetics, some arguing that it poses grave ethical, legal and social implications. It is precisely this type of research, they believe, that will lead to further discrimination and stigma for society's disadvantaged and underprivileged.

Behaviour genetics aside, almost each day it seems that one news account or another cites a research study claiming to have discovered the gene which produces this medical condition or is responsible for that affliction. Certain defective genes have been identified, such as those associated with severe genetic diseases as chronic granulomatous disease (CGD) or X-SCID (commonly referred to as 'baby in the bubble syndrome'). Such research, its proponents maintain, holds out the promise that gene therapy may lead to new forms of treatment for patients suffering from such conditions. Meanwhile some scientists are seeking to use new techniques to alter the genetic make-up of cell material from a human egg, the hope being that such an intervention will help prevent the eventual baby inheriting certain genetic diseases. Using IVF in conjunction with newer techniques, they are able to screen embryos for features which may make the potential baby a better 'match' for a sibling in desperate need of a transplant, such as bone marrow for example. In heated exchanges, advocates have insisted that one child is simply being used to help another, while critics have argued that 'breeding children for spare parts' is reprehensible, and that this type of 'Frankenstein medicine' should be made illegal at once.

Much of the media interest in biotechnology can be attributed to the ongoing findings being publicized by the **Human Genome Project**. The largest biological science project in history, the aims of the project have been aptly described by Glasner (2000: 131) as being a 'metaphorical search for the "holy grail", to write the "book of life" '. The term 'human genome' refers to the chemical code needed to build a human being. The 'blueprint' comprises DNA, a lone molecular chain of phosphate and sugar in the shape of a double-helix ladder connected by rungs, called bases. Each person's DNA is unique, except in the case of identical twins. Determining the precise structure of DNA, first discovered in 1953 by Watson and Crick (see Watson 2000), is the key objective of the Human Genome Project. Two rival groups of scientists – Celera Genomics (a private company) and National Human Genome Research Institute (a public consortium), respectively – announced that they had deciphered the hereditary script in June 2000. The human genome consists of two sets of 23 giant DNA molecules, or chromosomes, with each set – one inherited from each parent – containing more than 3 billion chemical units. Some scientists are pointing to what they perceive to be a host of possible benefits of this knowledge. Biologists, for example, expect to use knowledge about the human genome to develop an array of diagnostics and treatments tailored to individual patients. Some treatments will exploit the body's own mechanisms of self-repair. Researchers also hope to be able to identify genes implicated in diseases, including many cancers, and mental illness (schizophrenia and depression).

Possible dangers associated with this knowledge, however, revolve around the risk that it will be misused by those anxious to exploit it, not least for financial gain. Highly charged public debates are underway with regard to the fair and proper use of the information gleaned from a person's **DNA profile**. Some employers, for example, have joined voices in law enforcement, advocating its use as a screening device for job applicants. Moreover, genetic testing in the workplace, they argue, will afford employers the opportunity to make more informed decisions about their employees' safety and working conditions – but only at the expense of safeguards to their employees' privacy, critics point out. In the case of the insurance industry, such information might allow companies to evaluate applicants for life assurance, for example, by placing them in different 'risk categories' on the basis of their genetic predisposition to certain diseases. What might make financial sense for an insurance company anxious to reduce its 'exposure to risk' may, at the same time, leave some people unable to secure adequate insurance. Even at the level of medicine, critics warn that current racial and economic iniquities indicative of healthcare delivery could worsen in the event that expensive genetic technology takes over.

In all of the debates sparked by human genetic engineering science, however, one topic presently stands out as the most controversial in the eyes of the world's media. The prospect of human cloning becoming a possibility, even a probability, promises to become one of the most passionately debated issues in the years to come.

Human clones

It was in the early evening of a warm summer's day, 5 July 1996, that a rather unusual lamb was born in Scotland. Despite resembling any other baby lamb, 'Dolly' was unique, and destined to become the most famous lamb in history. As people living around the globe would soon discover to their astonishment on 24 February 1997, Dolly was a **clone** (the story had been broken the day before in the *Observer* newspaper). The worlds of science fiction and science fact had unexpectedly converged once again, and this time the implications – as voices on the front page of virtually every newspaper agreed – were particularly unsettling.

Medical breakthrough, or moral breakdown? Cloning was one feat of genetic engineering too far in the eyes of many scientists, few of whom had believed it would be achieved in their lifetime. Even leading researchers in the field of mammalian genetics and embryology were shocked by the news, some wondering if it was a cruel hoax. Others made the immediate jump in thought to wonder if the same techniques might be applied to humans. They were not alone – as Pence (1998: 1) writes, 'thirty hours after the news of Dolly hit the streets, legislator John Marchi announced a bill to make human cloning illegal in New York State'. In the words of Silver (1998), a biologist:

> Of course, it wasn't the cloning of a sheep that stirred the imagination of billions of people. It was the idea that humans could now be cloned as well in a manner akin to taking cuttings from a plant, and many people were terrified by the prospect. Ninety percent of Americans polled within the first week after the story broke felt that human cloning should be banned. And the opinions of many media pundits, ethicists, and policymakers, though not unanimous, seemed to follow those of the general public. The idea that humans might be cloned was called 'morally despicable', 'repugnant', 'totally inappropriate', as well as 'ethically wrong, socially misguided, and biologically mistaken'.
>
> (Silver 1998: 108)

Around the world a diverse array of institutional stakeholders, each articulating their respective perceptions of science and technology, sought to define its significance for humankind. Imagery from science fiction was drawn upon in order to tap into the public mood, frequently with great rhetorical effect.

Members of the public had been acquainted with the idea of scientists duplicating people long before the actual word 'clone' gained popular currency, due in part to its treatment in a number of different science fiction texts – even before *Brave New World* appeared. Closer to the time of the Dolly announcement, however, the possibility of human cloning had been explored in places such as Alvin Toffler's book *Future Shock* in 1970, and the 1973 Woody Allen film *Sleeper*. Allen's use of the concept for comedic effect was sharply counterpoised by Ira Levin's novel *The Boys from Brazil* a few years later. Published in 1976, and soon adapted into a film, the novel revolves around the cloning of tissue from Adolf Hitler's body by neo-Nazis exiled to Brazil. Their evil intent is to create an army of Hitlers, each sharing the dictator's genetic profile, to rule the world; 94 women are impregnated to act as surrogate mothers, and the subsequent offspring adopted by couples living in circumstances emulating those experienced by Hitler himself as a child. The experiment is foiled in the end, but not before providing the reader with a frightening vision of how cloning technology might one day be used to achieve sinister ends. In the years to follow, related novels to appear would include Nancy Freedman's *Joshua, Son of None*, depicting the cloning of US President John F. Kennedy. Of the more recent examples, several important ones in feminist science fiction standout, such as Fay Weldon's *The Cloning of Joanna May* (adapted for British television in 1991), and Anna Wilson's *Hatching Stones*.

Arguably the most influential of fictional representations of cloning prior to the Dolly announcement, albeit not of the human variety, is the Spielberg directed film *Jurassic Park*. Based on Michael Crichton's best-selling novel, the film was released in 1993 and promptly became the most successful film in worldwide box-office history for that time. Central to the story is the cloning of a series of different dinosaurs using fossilized DNA, the aim being to create attractions for 'the most advanced amusement park in the world' where members of the public will pay to watch them interact. To explain how this is achieved, an innovative filmic device is used. Specifically, a short animated film – titled *Mr DNA* – is presented within the film for the benefit of the central characters, as well as us, outlining the steps taken to clone a dinosaur. It sounds straightforward enough – ancient dinosaur DNA is recovered from mosquitoes embedded in fossilized amber for millions of years, their bodies containing blood that they had extracted from dinosaur skin prior to their death. The park's scientists use this DNA as a genetic 'blueprint' of sorts, splicing it together with frog DNA to fill in the missing code. The combined genetic material is then inserted into a crocodile or ostrich egg, depending on the species of dinosaur, which is then left to mature in the hatchery. Dinosaurs, long extinct, are brought back to life with a roar. 'The central modern myth of scientific resurrection, *Frankenstein*, is itself resurrected in the story of *Jurassic Park*,' as Mitchell (1998: 95) points out, 'in which the new technologies of genetic engineering, cloning, and computer science are mobilized to create a new kind of "animation" of the dead'. It soon transpires that the word 'animation' is something of an understatement, as courtesy of remarkable special effects, the remainder of the film is devoted to showing just how far beyond human control the dinosaurs can be ('Life escapes all barriers. Life breaks free,' states one character, Ian Malcolm). Each and every 'fail-safe' system designed to contain the creatures fails catastrophically. Chaos reigns, violently so, and it is corporate science that is held to blame. 'Genetic power is the most awesome force the planet's ever seen,' Malcolm exclaims, 'but you wield it like a kid that's found his dad's gun.' Later in the same scene he adds, 'Your scientists were so preoccupied with whether or not they could, they didn't bother to stop to think if they should.'

The social meanings of cloning articulated via *Jurassic Park*, it seems fair to say, will have had an impact – possibly in a profound way – on the images of genetics in public circulation at the same time. Crichton's story, as Nelkin and Lindee (1995: 54) observe, is a 'popular catechism promoting the idea of "immortal DNA" ', as well as being 'a morality play about forbidden fruit and the dangers of scientists playing God' (see also Franklin et al. 2000). For some individuals watching the film, particularly young people, it served as a primer of sorts, explaining in general terms what cloning entails while, at the same time, making a bold imaginative statement about its social consequences when control collapses. 'Imaginary tools', as Van Dijck (1998: 2–3) argues, 'are crucial assets in the dissemination of genetic knowledge, as they are used to shape this science's public face.' More than journalism, in her view, science fiction has 'formed the apparent playground where the cultural implications of cloning could be discussed, its social permissiveness be probed, and larger philosophical concerns about human integrity be raised' (Van Dijck 1998: 55). Particularly salient where tales of cloning are concerned, she suggests, are 'what if' scenarios; for example: 'what will happen to the human race if selected individuals can be infinitely reproduced?' (1998: 55). Compelling new questions, it follows, are generated about a life's 'natural' duration, or about what constitutes the 'essence' of an individual? Whether science fiction or journalism has a relatively greater influence on public perceptions of cloning in any one discursive instance is a matter of investigation, in my view, but it is readily apparent that both inform this wider cultural backdrop at any given moment.

In reading the news coverage of the Dolly announcement against these and related contours of this cultural backdrop, it is illuminating to document the ways in which factual discourses were shaped by fictional ones. Dolly was created using DNA from an adult sheep, something which had never been done successfully before. The researchers were led by Dr Ian Wilmut, an embryologist at the Roslin Institute near Edinburgh, working in collaboration with PPL Therapeutics. They took a mammary cell from a 6-year-old sheep and altered its DNA so as to make it acceptable to another sheep's egg (that it was a mammary cell led to the wry choice of the name Dolly, after country and western singer Dolly Parton). The egg's own genetic material was then removed and replaced by the DNA from the adult sheep by fusing the egg with the adult cell. The fused cells, carrying the adult DNA, grew and divided like they would in an egg fertilized the ordinary way, eventually forming an embryo. This embryo was then implanted into another ewe, which gave birth to Dolly in July 1996. A startling example of asexual reproduction, one that reportedly took 277 such fusions to achieve, Dolly had in effect two mothers but not a father.

Interestingly, when Wilmut was asked by a reporter at the time of the announcement whether Dolly should be considered to be 7 months old or 6 years old (the age of the cloned adult lamb), he was unable to say for certain. 'I can't answer that,' he admitted. 'We just don't know' (cited in Pence 1998: 18; see also Kolata 1997).

If prior to Dolly's arrival animal cloning had been the subject of scientific conjecture, now it was suddenly real. In the eyes of many of the journalists charged with explaining these types of details to their audiences, this was science gone mad. From the outset Dolly was characterized by some as a 'genetic mutant', and as such surely the first seed being sown in what would inevitably lead to the destruction of humankind. The prospect of 'carbon-copy human reproduction' generated uproarious public debates (the mad scientist in such scenarios becoming a 'mad Xeroxer'), with some of the voices participating bordering on the hysterical. Efforts to point out that identical twins are clones, albeit 'naturally achieved' ones, and that cloning methods might have benefits for those involved in IVF, were difficult to hear in the media furore about human cloning. Wilmut acknowledged at the time of the initial announcement that a similar technique might make human cloning theoretically possible, but had insisted that he found the very idea abhorrent. Indeed, it could be safely dismissed as the subject of science fiction, in his view – a position echoed, in turn, in a headline appearing in *The Guardian* that day: 'Scientists scorn sci-fi fears over sheep clone.' As Wilmut informed the press: 'We would find it ethically completely unacceptable and we would not do it.' Instead he was convinced that the breakthrough might lead, in the long term, to the development of animals that could produce drugs to treat human diseases, such as haemophilia, more cheaply than current methods (the secrecy around the breakthrough was due to a corresponding patent application). Similarly held open was the possibility that cloned animals might be able to serve as models for human diseases, such as cystic fibrosis, thereby allowing scientists to test new treatments.

Despite the reassurances of Wilmut and his colleagues that they would never attempt to clone humans, news coverage of the stunning breakthrough over the next few weeks continued to focus on precisely this issue. The flurry of publicity surrounding the controversy opened up the discursive terrain for a multitude of positions to be rehearsed. Huxford (2000), in his study of the newspaper reporting in Britain and the US over the ensuing month, suggests that a headline in London's *Daily Mail*, 'Monsters or a Miracle?', highlighted an opposition which set the tone for the 'avalanche' of copy that followed. Journalists, he argues, turned to certain science fiction frames that emphasized anti-science themes, in part by 'establishing a series of narrative oppositions through which the coverage might be shaped: science versus religion, high culture versus low culture, the Romantic sense of the individual versus mass society' (Huxford 2000: 187). Almost half (46 per cent) of the 204 articles surveyed made references to science fiction imagery, the vast majority in a negative manner by raising fears about the future use of cloning technology. 'British sheep clone raises spectre of Frankenstein', declared the Reuters news service (24 February 1997), while the main headline in the *News and Observer* in Raleigh, North Carolina on 9 March 1997 demanded to know: 'Are we cloning a Brave New World?' Significantly, Huxford's (2000) findings suggest that in those news items with science fiction associations, Huxley's novel was the most frequently work cited by far, followed by Shelley's *Frankenstein*. These two texts accounted for the bulk of the references, with 32 and 21 respectively. Next came *The Boys from Brazil* with 13, and then with 4 references the comedy film *Multiplicity* – its central character, played by Michael Keaton, is repeatedly cloned with each 'copy' being a progressively inferior version of the 'original' – which had been in cinemas the previous year. 'The use of science fiction allusions in the newspapers', Huxford's study suggests, 'had more to do with cueing certain cultural fears, as embodied in popular science fiction, than with providing an understanding of cloning' (Huxford 2000: 192).

Further evidence to support this line of argument is found in Priest's (2001) analysis of 'elite' or 'agenda-setting' US newspaper coverage of the controversy, namely from 1994 to its aftermath in 1997. Focusing primarily on news items published by *The New York Times*, the *Washington Post*, the *Los Angeles Times* and the *Wall Street Journal*, she argues that the Dolly story dominated the US media in a way that perhaps no other science story had ever done before. Of particular interest for her study was the extent to which the story became one about disagreements over cloning's ethical implications. 'Both the norms of professional journalistic practice and a particularly American cultural perspective contributed to the prominence of this story,' Priest (2001: 59) writes, 'creating a novel news "frame" in which the opinions of ethicists and occasionally of religious leaders acted as counterpoint to those of scientists.' Journalists, she points out, thrive on controversy. Cloning satisfied the needs of 'gee whiz' science reporting, but in her opinion the continuing salience of the issue had more to do with its role as a 'crystallizing symbol' for public concerns about biotechnology. That is to say, the controversy arose not over the technical facts or their interpretation *per se*, but rather over the ethics of making particular use of those facts. Ethicists' comments were regularly used by journalists to 'balance' scientific points of view, the effect of which – however unintentionally on the part of the reporters involved – was to contain the debate in highly circumscribed ways. The cloning controversy, Priest argues, 'was essentially harmless to the status quo arrangement whereby biotechnological development is driven primarily by narrow institutional economic interests rather than issues of broader social benefit' (Priest 2001: 67). As a result, she suggests, the pre-existing power relationships characteristic of the area were largely allowed to stand unchallenged. Indeed, when looking across the time period under scrutiny, her study indicates that 'objections to genetic engineering for agricultural or pharmaceutical (rather than human reproductive) purposes on economic, environmental, or other grounds were nearly invisible' (Priest 2001: 67). The consequences for how the ensuing public debate was structured were thus far-reaching (this conclusion strikes a resonance in Neresini's (2000) analysis of the Italian coverage).

It is clearly the case that as the issues surrounding biotechnology become evermore pressing, new and more ideologically diverse spaces need to be identified and sustained across the media field for public deliberations about their implications. Informing this intervention to enrich the spaces available for dialogue needs to be a greater recognition of the ways in which fictional and factual representations conjoin, intersect and contradict one another. As Gaskell et al. (1998) observe in their discussion of biotechnology, 'the importance of individual and collective imagination in shaping public perceptions can scarcely be exaggerated.' Indeed, as they proceed to add, the 'cultural resonance of key phrases – "test tube baby", "genetic engineering", "cloning", etc. – has as much to do with their metaphorical powers as with their scientific and technological significance' (Gaskell et al. 1998: 10). The precise inflection of these and related phrases varies dramatically, of course, depending on the economic, political and ideological interests at stake. For some advocates of biotechnology, for example, there are exciting advances to be gained by therapeutic cloning, that is, where scientists create human embryos in the laboratory and then draw from them special cells for use in medical treatments. Typically cited in this context are beneficial possibilities for transplants as donated cells 'grown to order' are less likely to be rejected by a patient's immune system because they are recognized as the body's own cells. This type of cloning might make transplants of organs, such as kidneys and hearts, standard treatment. It may also allow skin to be made for burns victims, or pancreatic cells to produce insulin for diabetics, and so forth. Advocates similarly point to potentially revolutionary treatments for leukaemia, Parkinson's and Alzheimer's disease, and other degenerative diseases.

Much of the criticism of human cloning techniques, in contrast, revolves around the scientists' use of human **embryonic stem cells** (parent cells for all tissues in body) which, as their name suggests, can be obtained only from embryos. Adult stem cells may be used in some circumstances, but are believed to lack the complete flexibility of embryonic cells. Various religious and anti-abortion groups have been particularly fierce in condemning the use of embryonic stem cells as an aberration of normalcy. Many of them argue that an embryo (including the small clump of cells being 'harvested', to use their word) is a unique human life and as such is sacrosanct. Fearful that scientists are seeking to 'play God' or 'defy nature', they have directed intense criticism at those conducting research on human cloning. One such individual is Italian physician and fertility specialist, Professor Severino Antinori (his use of IVF to help a 62-year-old woman become a mother had created a media furore in the mid-1990s). On several occasions, Antinori has publicly declared his commitment to develop human cloning (or, as he prefers to say, 'genetically reprogramming') techniques in order to help infertile couples have a child. This interview exchange between Antinori and Tim Adams (2001), a journalist, is telling:

> 'People fear that we lose individuality,' Antinori says. 'But people are shaped by a million different things! By the environment, by their times. I know two "clones", identical twins. One is a doctor, one is whatever. You cannot tell them apart! But they could not be more different. What is to fear?'
>
> Won't the technology inevitably fall into the wrong hands?
>
> 'People said the same about the atomic bomb,' he says.
>
> Well, exactly. . .
>
> 'But it has not happened!' he says.
>
> Won't people see it as a short cut to immortality? To recreate themselves?
>
> 'Imagine,' he says, looking at me, grinning, 'we could create a thousand Tims! Who would want that? *Terrible!* A thousand Tims! *No Way!* But that is what regulation is for, not to stop science, to make sure it is used positively.'
>
> (cited in Adams 2001; emphasis in original)

His stance, obviously very much a maverick one, has gained some, albeit limited support in the scientific community. For those sharing Antinori's vision, cloning techniques promise to bring the joys of parenthood to individuals otherwise denied the opportunity to have a genetically related child. Much is made of how the public uproar over previous developments – blood transfusions, organ transplants and, of course, IVF – eventually subsided as the potential benefits became more widely recognized. This will be the case with cloning as well, they contend, but in the meantime they believe it is important that such work continues, not least to help secure a reliable source of embryonic stem cells. Therapeutic cloning, they hasten to point out, is not the same as reproductive cloning (the latter, in any event, being illegal in many countries). For those holding diametrically opposed views, however, this distinction is ultimately untenable. Indeed, many of them fear that it is the first step down a slippery slope toward 'designer children' and human clones being born en masse. Some of these counter-arguments are made on philosophical grounds, others on the conviction that it is not safe. Cloning experiments, they maintain, will lead to babies suffering abnormalities, or dying soon after birth – and here they often point to studies of cloned mice, cows, pigs and goats for comparative evidence. The dream of 'disease-free super children', in their view, is set to become a tragic nightmare.

To close, we may return to the question: what does it mean to be human? A definitive answer continues to prove elusive, as will always be the case. Indeed, as so many of the respective voices in this chapter's discussion demonstrate, precisely what it means to be fully human will vary dramatically from one speaker's vantage point to the next. Moreover, it is readily apparent that an array of alternative definitions are in circulation across the media field, each mobilizing its preferred formulation of science in order to legitimize the respective future for humankind it projects. Wherein lies some agreement, however, is that this question has never mattered quite so much as it does today. Still, is it fair to suggest that the era of the posthuman beckons, and if so can we afford to be optimistic? Hayles (1999), in her exploration of embodiment in an information age, assumes a radical stance in arguing that it is just possible that the posthuman is to be welcomed and embraced, rather than feared and abhorred. In either case, though, she contends:

> The best possible time to contest for what the posthuman means is now, before the trains of thought it embodies have been laid down so firmly that it would take dynamite to change them. Although some current versions of the posthuman point toward the anti-human and the apocalyptic, we can craft others that will be conducive to the long-range survival of humans and of the other life-forms, biological and artificial, with whom we share the planet and ourselves.
>
> (Hayles 1999: 291)

The exigent need to examine afresh the complexities, and lived contradictions, of 'what it means to be human' becomes more urgent each day. Thus Hayles's emphasis on the importance of finding new ways to rethink the fluidly contingent dynamics by which such definitions are changing is for this chapter – as well as for this book, I believe – an appropriately provocative challenge to end on.

Further reading

Gray, C.H. (2001) *Cyborg Citizen*. New York: Routledge.

Haraway, D.J. (1991) *Simians, Cyborgs, and Women*. London: Free Association Books.

Hartouni, V. (1997) *Cultural Conceptions: On Reproductive Technologies and the Remaking of Life*. Minneapolis, MN: University of Minnesota Press.

Hayles, N.K. (1999) *How We Became Posthuman*. Chicago: University of Chicago Press.

Nelkin, D. and Lindee, M.S. (1995) *The DNA Mystique*. New York: W.H. Freeman.

Nottingham, S. (2000) *Screening DNA: Exploring the Cinema–Genetics Interface*. Stevenage: DNA Books.

Turney, F. (1998) *Frankenstein's Footsteps: Science, Genetics and Popular Culture*. New Haven, CT: Yale University Press.

Van Dijck, J. (1998) *Imagenation: Popular Images of Genetics*. London: Macmillan.

Watson, J.D. (2000) *A Passion for DNA*. Oxford: Oxford University Press.

GLOSSARY

AI: an abbreviation for artificial intelligence, whereby human processes are modelled by computers. It is generally agreed that for a computer to actually possess AI, it would have to be self-aware – a development yet to be achieved.

AIDS: acquired immune deficiency syndrome, the name given to a range of opportunistic infections and malignancies arising due to the body's immune system having been compromised after infection with **HIV**.

Android: a **robot** possessing a human appearance.

Biotechnology: this term, as its root words 'bio' and 'technology' suggest, generally refers to the application of biological systems and processes in the production of a material entity, typically for commercial, medical or nutritional reasons. Current uses of the term tend to emphasize the application of cellular and molecular processes, involving **DNA** and proteins, in making products.

BSE: bovine spongiform encephalopathy is a neurological – and ultimately fatal – disorder of cattle, believed to be caused by **prions**. Tiny lesions in the brain engender behavioural changes, including loss of muscle control and co-ordination (popularly known as 'mad cow disease').

Cell: the basic structural and functional unit of which organisms consist.

CJD: Creutzfeldt–Jakob disease is a rare and fatal neurological disorder of humans caused by **prions**. A variant of the disease is believed to be linked to **BSE**. Named after Hans Gerhard Creutzfeldt and Alfons Maria Jakob.

Clone: an identical copy of an individual; more specifically, a clone is a group of genetically identical **cells** or organisms produced asexually from an original cell or organism.

Cyberpunk: a subgenre of science fiction writing, emergent in the 1980s, where the future is typically depicted as being overdetermined by technological processes and information networks. The 'punk' component of the term signals its subversive intent, indicative of an alienated, counterculture politics. The term was popularized, in part, with reference to the social environment portrayed by William Gibson in his 1984 novel *Neuromancer*.

Cyborg: a contraction of 'cybernetic organism', and as such the product of the merging of human and machine elements to produce a hybrid. Sometimes used to refer to people who have been modified mechanically, either to perform specific tasks or to function in an alternative environment.

DNA: deoxyribonucleic acid, a substance present in nearly all living organisms. It is genetic material made up of a chain of individual units called nucleotides (each nucleotide consisting of a base joined to a sugar and a phosphate group). DNA molecules carry the genetic information necessary for **cells** to operate, and also control the inheritance of characteristics. J.D. Watson and F.H.C. Crick, using X-ray crystallography, proposed the structure of DNA in 1953 (see Watson 2000).

DNA profiling: the analysis of **DNA**, using samples of bodily tissues or fluids, in order to identify an individual through patterns in the genetic material; sometimes also called genetic fingerprinting. The acquired information may be used for a number of purposes, including as forensic evidence, to determine genetic relatedness, or diagnostically to determine susceptibility to genetically inherited disorders.

E. coli: the bacterium Escherichia coli, some types of which cause forms of food poisoning.

Ecology: the study of organisms in relation to one another, and to their environmental surroundings.

Embryo: an unborn offspring; in human terms, it refers to the first eight weeks of existence from conception.

Embryonic stem cells: undifferentiated **cells** possessing the ability to form any adult cell in the human body. They are derived from fertilized **embryos** less than a week old. In their undifferentiated state in the laboratory, stem cells show an ability to divide indefinitely. Some scientists hope that these cells will provide an unlimited source of clinically important adult cells, such as bone, muscle, liver or blood cells, that can be used to treat a host of cell-based diseases.

Foot-and-mouth-disease: a highly contagious viral disease of animals such as cattle and swine (called hoof-and-mouth-disease in some countries). It also affects sheep, goats, deer and other cloven-hoofed ruminants. Although only rarely fatal, the disease leaves animals debilitated for up to six months, causing economic hardship for farmers whose livelihood depends on the production of meat and milk.

Gene: the fundamental unit of inheritance and function in a **cell**, usually identified with lengths of **DNA**. A gene is transmitted from parent to offspring, ordinarily as part of a chromosome. Gene splicing is where genetic material is modified by cutting a DNA molecule(s) and rejoining the cut ends to form a new synthesis.

Gene therapy: the insertion of functional genes into an **embryo** or into mature **cells** in place of missing or defective ones in order to correct genetic disorders.

Genetic engineering: the modification or manipulation of genetic material whereby **DNA** fragments from different sources are combined to make recombinant

DNA. This recombinant DNA may be inserted, in turn, into **cells** so as to alter their function. Sheep's milk may be made to produce human hormones, for example, or crops modified to make them disease resistant.

Genome: the totality of the genetic material of a **cell** or organism.

Global warming: a theory in climatology whereby it is believed the mean temperature of the Earth's atmosphere is rising due to the influence of the **greenhouse effect**. A significant rise in world temperatures would have potentially catastrophic climatic and environmental effects.

GM crops and food: genetically modified (or genetically manipulated) crops or food. See **genetic engineering**.

Herbicide: a substance toxic to plants, used to destroy weeds or unwanted vegetation.

HIV: the human immunodeficiency virus believed to cause **AIDS**. The virus infects T cells of the immune system, undermining the body's ability to ward off infection. HIV can be transmitted from person to person by contact with infected blood, semen or vaginal fluid, or from mother to foetus via the placenta. A person is said to be HIV positive in the event that antibodies to HIV are detected in their blood.

Human Genome Project: an international effort to identify the sequence of bases in the **DNA** that comprises the human **genome**. Already the project is providing an enhanced understanding of genetic processes, especially where diseases are concerned.

Insecticide: a compound or mixture that kills insects.

IVF: *in vitro* (Latin for 'in glass') fertilization is a technique in which gametes (germ cells) are mixed in a liquid in a dish to achieve fusion of eggs and sperm. The fertilized egg may be then implanted into the uterus.

Mad cow disease: see **BSE**.

Meteorite: a rocky meteoroid of sufficient size that it does not burn up completely upon entering the Earth's atmosphere, thereby reaching the planet's surface.

Nobel Prize: any of the six highly prestigious awards for academic study (present categories are chemistry, physics, physiology/medicine, literature, economics and peace). The annual prizes, adjudicated by Swedish learned societies, are named after Alfred Bernhard Nobel.

Nuclearism: refers to the pro-nuclear assumptions underpinning claims advanced to legitimize, in ideological terms, the continued use of nuclear technology, whether in the case of nuclear energy or nuclear weapons (see also Irwin et al. 2000).

Ozone layer: the layer of ozone (one of the allotropes of oxygen) in the Earth's upper atmosphere that protects life on the planet from most of the harmful effects of ultraviolet radiation from the sun. The depletion of this layer's concentration – at worst leading to an ozone hole – is thought to be caused by atmospheric pollution derived from the release of certain chemicals, such as chlorofluorocarbons (CFCs). Increased ultraviolet life is harmful, and is also regarded as a contributory factor to **global warming**.

Pathogen: any agent responsible for causing a disease.

Precautionary principle: this principle holds that action should be taken beforehand, possibly on the basis of suggestive rather than definitive evidence, to either reduce or avoid **risk**.

Prion: a protein particle, possibly derived from a degenerate virus, thought to act as a disease-causing agent. It has been implicated in various neurological disorders, such as scrapie in sheep, **BSE** in cows, and **CJD** in humans.

Pseudo-science: a set of beliefs or claims which adopts scientific terminology while, at the same time, violating the basic tenets of scientific principles.

Risk: the chance or possibility of danger (harm, loss, injury and so forth) or other adverse consequences actually happening. 'This concept', according to Beck (2000: xii), 'refers to those practices and methods by which the future consequences of individual and institutional decisions are controlled in the present.'

Risk society: a term coined by social theorist Ulrich Beck (1992a) to characterize the ways in which processes of 'reflexive modernization' are transforming societies. Far from simply arguing that modern societies are riskier than past societies, Beck contends that there has been a qualitative shift in the way people are living with risk. Briefly, their awareness of risk in their everyday lives is intensifying, a corresponding increase in the levels of uncertainty surrounding risk is leading to a greater dependence on experts to manage or control risk, and yet public trust in such expertise is being eroded. Science, as a result, no longer enjoys the public esteem it once did, nor is it perceived as having a monopoly on truth (see also Macnaghten and Urry 1998; McGuigan 1999; Adam et al. 2000).

Robot: a mechanical device capable of moving itself, and able to carry out a complex series of actions (including the manipulation of objects) automatically. Robotics is the study of robots, including the science of their design, construction, operation and application.

Salmonella: a bacterium of the genus *Salmonella*, infection of which can cause food poisoning.

Science wars: an ongoing, and at times acrimonious, debate between scientists and social scientists over the nature, value, meaning, legitimacy and authority of science in society (see Ross 1996; Labinger and Collins 2001). The debate was given a new impetus following the so-called **Sokal affair**.

Scientific literacy: the 'scientifically literate' adult, according to POST (1995), is someone who: 'a) has a basic vocabulary of scientific terms and concepts adequate to read reports of scientific disputes; b) distinguishes between science and **pseudo-science**; c) knows how science affects our daily lives.'

Sokal affair: in May 1996, physicist Alan Sokal's ostensibly serious article on 'quantum gravity', actually written as a parody of how some philosophers and cultural theorists discuss science, was published in the academic journal *Social Text*. He then revealed his hoax to the world, the ensuing uproar generating widespread public and media attention (see Editors, Lingua Franca 2000).

Stakeholder: any individual or group of people seeking to gain public support and/or policy-maker support for their particular position on an issue. They have a stake or interest in how a given decision, activity or claim is framed or understood, and what action ensues (or not) as a result.

Virtual reality: a computer-generated depiction of reality, which appears to be 'real' or at least 'realistic' for the person who 'enters' and 'interacts' with it.

Virus: a parasitic genetic element capable of replicating in **cells** and forming infectious particles, or remaining dormant in the cell. While a virus can be regarded as being biological in nature, it is not capable of the independent maintenance of life processes.

REFERENCES

Ackerman, F.J. (1997) *World of Science Fiction*. London: Aurum.

Adam, B. (1998) *Timescapes of Modernity: The Environment and Invisible Hazards*. London: Routledge.

Adam, B. (2000) The media timescapes of BSE news, in S. Allan, B. Adam and C. Carter (eds) *Environmental Risks and the Media*. London and New York: Routledge.

Adam, B., Beck, U. and van Loon, J. (eds) (2000) *The Risk Society and Beyond: Critical Issues for Social Theory*. London: Sage.

Adams, J. (1999) Cars, cholera, cows, and contaminated land: virtual risk and the management of uncertainty, in R. Bate (ed.) *What Risk? Science, Politics and Public Health*. Oxford: Butterworth Heinemann.

Adams, T. (2001) The clone arranger, *Observer (Review section)*, 2 December.

Adams, W.C. (1986) Whose lives count? TV coverage of natural disasters, *Journal of Communication*, 36(2): 113–22.

Aldiss, B. (1973) *Billion Year Spree*. London: Weidenfeld.

Alkon, P.K. (1994) *Science Fiction before 1900*. New York: Twayne.

Allan, S. (1999) *News Culture*. Buckingham and Philadelphia, PA: Open University Press.

Allan, S., Adam, B. and Carter, C. (eds) (2000) *Environmental Risks and the Media*. London and New York: Routledge.

Altman, D. (1986) *AIDS and the New Puritanism*. London: Pluto.

Altman, L.K. (2001a) The AIDS questions that linger, *New York Times*, 30 January.

Altman, L.K. (2001b) H.I.V. 'explosion' seen in East Europe and Central Asia, *New York Times*, 29 November.

Alwood, E. (1996) *Straight News: Gays, Lesbians and the News Media*. New York: Columbia University Press.

Anderson, A. (1997) *Media, Culture and the Environment*. London: UCL Press.

Anderson, A. (2000) Environmental pressure politics and the 'risk society', in S. Allan, B. Adam and C. Carter (eds) *Environmental Risks and the Media*. London and New York: Routledge.
Angenot, M. (1979) Jules Verne: the last happy utopianist, in P. Parrinder (ed.) *Science Fiction: A Critical Guide*. London: Longman.
Appleyard, B. (1992) *Understanding the Present*. London: Pan.
Ashley, M. (2000) *The Time Machines*. Liverpool: Liverpool University Press.
Atkinson, K. and Middlehurst, R. (1995) Representing AIDS: the textual politics of health discourse, in B. Adam and S. Allan (eds) *Theorizing Culture*. London: UCL Press; New York: NYU Press.
Baker, R. (2001) *Fragile Science*. London: Macmillan.
Bakir, T.V. (2001) Media agenda-building battles between Greenpeace and Shell: a rhetorical and discursive approach. Unpublished PhD thesis, University of Hull.
Barnard, P. (1999) *We Interrupt This Programme*. London: BBC.
Barr, J. and Birke, L. (1998) *Common Science?* Bloomington, IN: Indiana University Press.
Baudrillard, J. (1994) *Simulacra and Simulation*. Ann Arbor, MI: University of Michigan Press.
Beardsworth, A. and Keil, T. (1997) *Sociology on the Menu*. London: Routledge.
Beck, U. (1992a) *Risk Society: Towards a New Modernity*. London: Sage.
Beck, U. (1992b) From industrial society to risk society: questions of survival, social structure and ecological enlightenment, *Theory, Culture and Society*, 9: 97–123.
Beck, U. (1995) *Ecological Politics in an Age of Risk*. Cambridge: Polity.
Beck, U. (1996a) Risk society and the provident state, in S. Lash, B. Szerszynski and B. Wynne (eds) (1996) *Risk, Environment and Modernity: Towards a New Ecology*. London: Sage.
Beck, U. (1996b) World risk society as cosmopolitan society? Ecological questions in a framework of manufactured uncertainties, *Theory, Culture and Society*, 13(4): 1–32.
Beck, U. (1998) Politics of risk society, in J. Franklin (ed.) *The Politics of Risk Society*. Cambridge: Polity.
Beck, U. (2000) Foreword, in S. Allan, B. Adam and C. Carter (eds) *Environmental Risks and the Media*. London and New York: Routledge.
Becker, H.S. (1967) Whose side are we on? *Social Problems*, 14(3): 239–47.
Beder, S. (1997) *Global Spin: The Corporate Assault on Environmentalism*. Totnes, Devon and White River Junction, VT: Green Books and Chelsea Green Publishing.
Bell, A. (1994) Climate of opinion: public and media discourse on the global environment, *Discourse and Society*, 5(1): 33–64.
Bellon, J. (1999) The strange discourse of *The X-Files*, *Critical Studies in Mass Communication*, 16: 136–54.
Bennett, T. (1995) *The Birth of the Museum*. London: Routledge.
Best, S. and Kellner, D. (2001) *The Postmodern Adventure*. London: Routledge.
Beynon, J. (2002) *Masculinities and Culture*. Buckingham: Open University Press.

Bird, S.E. (1996) CJ's revenge: media, folklore, and the cultural construction of AIDS, *Critical Studies in Mass Communication*, 13: 44–58.

Blum, D. (1997) Investigative science journalism, in D. Blum and M. Knudson (eds) *A Field Guide for Science Writers*. New York: Oxford University Press.

Blum, D. and Knudson, M. (1997) Editors' note, in D. Blum and M. Knudson (eds) *A Field Guide for Science Writers*. New York: Oxford University Press.

Bodmer, W. (1985) *The Public Understanding of Science*. London: Royal Society.

Bodmer, W. and Wilkins, J. (1992) Research to improve public understanding programmes, *Public Understanding of Science*, 1: 7–10.

Boyer, P. (1985) *By the Bomb's Early Light: American Thought and Culture at the Dawn of the Atomic Age*. New York: Pantheon.

Bradburne, J.M. (1998) Dinosaurs and white elephants: the science center in the twenty-first century, *Public Understanding of Science*, 7: 237–53.

Brassley, P. (2001) The minister and the malady, *History Today*, 51(11): 26–8.

Brody, J.E. (2001) Clean cutting boards are not enough, *New York Times*, 30 January.

Broks, P. (1996) *Media Science before the Great War*. London: Macmillan.

Brookes, R. (2000) Tabloidization, media panics, and mad cow disease, in C. Sparks and J. Tulloch (eds) *Tabloid Tales*. Lanham, MD: Rowman and Littlefield.

Brookes, R. and Holbrook, B. (1998) 'Mad cows and Englishmen': gender implications of news reporting on the British beef crisis, in C. Carter, G. Branston and S. Allan (eds) *News, Gender and Power*. London and New York: Routledge.

Bucchi, M. (1998) *Science and the Media: Alternative Routes in Scientific Communication*. London: Routledge.

Bunyard, P. (1988) Nuclear energy after Chernobyl, in E. Goldsmith and N. Hildyard (eds) *The Earth Report: Monitoring the Battle for our Environment*. London: Beasley.

Campbell, J.E. (2001) Alien(ating) ideology and the American media, *International Journal of Cultural Studies*, 4(3): 327–47.

Canaday, J. (2000) *The Nuclear Muse*. Madison, WI: University of Wisconsin Press.

Caplan, P. (2000) 'Eating British beef with confidence': a consideration of consumers' responses to BSE in Britain, in P. Caplan (ed.) *Risk Revisited*. London: Pluto.

Carter, D. (1988) *The Final Frontier*. London: Verso.

Carter, S. (1997) Reducing AIDS risk: a case of mistaken identity?, *Science as Culture*, 6(2): 220–45.

Cartmell, D., Hunter, I.Q., Kaye, H. and Whelehan, I. (eds) (1999) *Alien Identities*. London: Pluto.

Chapman, G., Kumar, K., Fraser, C. and Gaber, I. (1997) *Environmentalism and the Mass Media: The North–South Divide*. London: Routledge.

Chibnall, S. (1999) Alien women: the politics of sexual difference in British sf pulp cinema, in I.Q. Hunter (ed.) *British Science Fiction Cinema*. London: Routledge.

Clarke, A.C. (1993) Introduction, in H.G. Wells, *The War of the Worlds*. London: Everyman.

Clute, J. and Nicholls, P. (1999) *The Encyclopedia of Science Fiction*. London: Orbit.
Cohen, S. (1972) *Folk Devils and Moral Panics*. London: MacGibbon and Kee.
Cohn, V. (1997) Coping with statistics, in D. Blum and M. Knudson (eds) *A Field Guide for Science Writers*. New York: Oxford University Press.
Collins, H. and Pinch, T. (1998) *The Golem at Large*. Cambridge: Cambridge University Press.
Cook, J. (1999) Adapting telefantasy: the *Doctor Who and the Daleks* films, in I.Q. Hunter (ed.) *British Science Fiction Cinema*. London: Routledge.
Cooter, R. and Pumfrey, S. (1994) Separate spheres and public places: reflections on the history of science popularization and science in popular culture, *History of Science*, 32: 237–67.
Cottle, S. (1998) Ulrich Beck, 'risk society' and the media: a catastrophic view?, *European Journal of Communication*, 13(1): 5–32.
Cottle, S. (2000) TV news, lay voices and the visualisation of environmental risks, in S. Allan, B. Adam and C. Carter (eds) *Environmental Risks and the Media*. London and New York: Routledge.
Coupland, J. and Coupland, N. (2000) Selling control: ideological dilemmas of sun, tanning, risk and leisure, in S. Allan, B. Adam and C. Carter (eds) *Environmental Risks and the Media*. London and New York: Routledge.
Crimp, D. (1993) Accommodating Magic, in M. Gerber, J. Matlock and R.L. Walkowitz (eds) *Media Spectacles*. New York: Routledge.
Critcher, C. (in press) *Moral Panics and the Media*. Buckingham: Open University Press.
Cunningham, A.M. (1986) Not just another day in the newsroom: the accident at TMI, in S.M. Friedman, S. Dunwoody and C.L. Rogers (eds) *Scientists and Journalists*. New York and London: Free Press.
Daley, P. and O'Neill, D. (1991) 'Sad is too mild a word': press coverage of the *Exxon Valdez* oil spill, *Journal of Communication*, 41(4): 42–57.
Daley, S. (2001) French report faults response to mad cow crisis, *New York Times*, 18 May.
Darier, E. (ed.) (1999) *Discourses of the Environment*. Oxford: Blackwell.
Dawkins, R. (1998) *Unweaving the Rainbow*. Harmondsworth: Penguin.
Dean, J. (1998) *Aliens in America: Conspiracy Cultures from Outerspace to Cyberspace*. Ithaca, NY: Cornell University Press.
DeFrancesco, L. (2001) Quickening the diagnosis of mad cow disease, *The Scientist*, 15(12): 22.
Detjen, J. (1997) Environmental writing, in D. Blum and M. Knudson (eds) *A Field Guide for Science Writers*. New York: Oxford University Press.
Dickinson, R. (1990) Beyond the moral panic: Aids, the mass media and mass communication research, *Communications*, 15(1/2): 21–36.
Dickson, D. (2000) Science and its public: the need for a third way, *Social Studies of Science*, 30(6): 917–23.
Disch, T.M. (1998) *The Dreams Our Stuff is Made of*. New York: Touchstone.
Dornan, C. (1988) The 'problem' of science and the media: a few seminal texts in their context, 1956–1965, *Journal of Communication Inquiry*, 12(2): 53–70.

Dornan, C. (1990) Some problems in conceptualizing the issue of 'science and the media', *Critical Studies in Mass Communication*, 7: 48–71.

Dubois, L. (1996) A spoonful of blood: Haitians, racism and AIDS, *Science as Culture*, 6: 7–43.

Dunbar, R. (1995) *The Trouble with Science*. London: Faber and Faber.

Durant, J. (1992a) Introduction, in J. Durant (ed.) *Museums and the Public Understanding of Science*. London: Science Museum.

Durant, J. (1992b) Editorial, *Public Understanding of Science*, 1: 1–5.

Durant, J. (1993) What is scientific literacy?, in J. Durant and J. Gregory (eds) *Science and Culture in Europe*. London: Science Museum.

Durant, J. (1998a) Pseudo-science, total fiction, *The Independent*, 21 August.

Durant, J. (1998b) Once the men in white coats held the promise of a better future. . ., in J. Franklin (ed.) *The Politics of Risk Society*. Cambridge: Polity.

Durant, J., Bauer, M.W. and Gaskell, G. (eds) (1998) *Biotechnology in the Public Sphere*. London: Science Museum.

Dutt, B. and Garg, K.C. (2000) An overview of science and technology coverage in Indian English-language dailies, *Public Understanding of Science*, 9: 123–40.

Editors, Lingua Franca (2000) *The Sokal Hoax*. Lincoln, NE: University of Nebraska Press.

Eldridge, J. (1999) Risk, society and the media: now you see it, now you don't, in G. Philo (ed.) *Message Received*. London: Longman.

Elena, A. (1997) Skirts in the lab: *Madame Curie* and the image of the woman scientist in the feature film, *Public Understanding of Science*, 6: 269–78.

Epstein, S. (1996) *Impure Science: AIDS, Activism and the Politics of Knowledge*. Berkeley, CA: University of California Press.

Evans, A.B. (1999) The origins of science fiction criticism: from Kepler to Wells, *Science Fiction Studies*, 26(2): 163–86.

Farmelo, G. (1998) Book review, *Public Understanding of Science*, 7: 352–4.

Fauvel, J. (1989) AIDS culture, *Science as Culture*, 7: 43–68.

Featherstone, M. and Burrows, R. (eds) *Cyberspace/Cyberbodies/Cyberpunk*. London: Sage.

Fishman, M. (1980) *Manufacturing the News*. Austin, TX: University of Texas Press.

Flatow, I. (1997) Magazine style, in D. Blum and M. Knudson (eds) *A Field Guide for Science Writers*. New York: Oxford University Press.

Ford, B.J. (2000) *The Future of Food*. London: Thames and Hudson.

Franklin, J. (ed.) (1998) *The Politics of Risk Society*. Cambridge: Polity.

Franklin, S., Lury, C. and Stacey, J. (2000) *Global Nature, Global Culture*. London: Sage.

Friedman, S. (1981) Blueprint for breakdown: Three Mile Island and the media before the accident, *Journal of Communication* 31(2): 116–28.

Friedman, S.M. (1986) The journalist's world, in S.M. Friedman, S. Dunwoody and C.L. Rogers (eds) *Scientists and Journalists*. New York: Free Press.

Friedman, S.M., Dunwoody, S. and Rogers, C.L. (eds) (1986) *Scientists and Journalists*. Washington, DC: AAAS.

Friedman, S.M., Gorney, C.M. and Egold, B.P. (1987) Reporting on radiation: a content analysis of Chernobyl coverage, *Journal of Communication*, 37(3): 58–78.

Friedman, S.M., Dunwoody, S. and Rogers, C.L. (eds) (1999) *Communicating Uncertainty*. Mahwah, NJ: Lawrence Erlbaum.

Fuller, S. (1997) *Science*. Buckingham: Open University Press.

Gaber, I. (2000) The greening of the public, politics and the press, 1985–1999, in J. Smith (ed.) *The Daily Globe: Environmental Change, the Public and the Media*. London: Earthscan.

Gaddy, G.D. and Tanjong, E. (1986) Earthquake coverage by the Western press, *Journal of Communication*, 26(2): 105–12.

Garfield, S. (2001) AIDS: the first 20 years, *The Guardian*, 3 June.

Garrett, L. (1997) Covering infectious diseases, in D. Blum and M. Knudson (eds) *A Field Guide for Science Writers*. New York: Oxford University Press.

Gascoigne, T. and Metcalfe, J. (1997) Incentives and impediments to scientists communicating through the media, *Science Communication*, 18(3): 265–82.

Gaskell, G., Bauer, M.W. and Durant, J. (1998) The representation of biotechnology, in J. Durant, M.W. Bauer and G. Gaskell (eds) *Biotechnology in the Public Sphere*. London: Science Museum.

Gauntlett, D. (1996) *Video Critical: Children, the Environment and Media Power*. Luton: University of Luton Press/John Libbey.

Gernsback, H. (1926) A new sort of magazine, *Amazing Stories*, 1(1): 3.

Giddens, A. (1998) Risk society: the context of British politics, in J. Franklin (ed.) *The Politics of Risk Society*. Cambridge: Polity.

Glasner, P. (2000) Reporting risks: problematising public participation and the Human Genome Project, in S. Allan, B. Adam and C. Carter (eds) *Environmental Risks and the Media*. London and New York: Routledge.

Goldblatt, D. (1996) *Social Theory and the Environment*. Cambridge: Polity.

Goldsmith, M. (1986) *The Science Critic*. London: Routledge and Kegan Paul.

Goldstein, R. (1991) The implicated and the immune: responses to AIDS in the arts and popular culture, in D. Nelkin, D.P. Willis and S.V. Parris (eds) *A Disease of Society*. Cambridge: Cambridge University Press.

Gould, S.J. (1996) *Dinosaur in a Haystack*. London: Jonathan Cape.

Gray, C.H. (2001) *Cyborg Citizen*. New York: Routledge.

Greenberg, J. (1997) Using sources, in D. Blum and M. Knudson (eds) *A Field Guide for Science Writers*. New York: Oxford University Press.

Greenberg, M.R., Sachsman, D.B., Sandman, P.M. and Salomone, K.L. (1989) Risk, drama and geography in coverage of environmental risk by network TV, *Journalism Quarterly*, 66(2): 267–76.

Gregory, C. (2000) *Star Trek: Parallel Narratives*. London: Macmillan.

Gregory, J. and Miller, S. (1998) *Science in Public*. Cambridge: Perseus.

Gregory, J. and Miller, S. (2001) Caught in the crossfire? The public's role in the science wars, in J.A. Labinger and H. Collins (eds) *The One Culture?* Chicago: University of Chicago Press.

Grimshaw, J. (1997) The nature of AIDS-related discrimination, in J. Oppenheimer and H. Reckitt (eds) *Acting on AIDS*. London: Serpent's Tail.

Gross, E. and Altman, M.A. (1995) *Captains' Logs*. Boston, MA: Little, Brown.
Grove-White, R. (1998) Risk society, politics and BSE, in J. Franklin (ed.) *The Politics of Risk Society*. Oxford: Polity.
Guedes, O. (2000) Environmental issues in the Brazillian press, *Gazette*, 62(6): 537–54.
Hackman, W. (1992) 'Wonders in one closet shut': the educational potential of history of science museums, in J. Durant (ed.) *Museums and the Public Understanding of Science*. London: Science Museum.
Haining, P. (1983) *Doctor Who: A Celebration*. London: Virgin.
Hall, S. (1981) The determinations of news photographs, in S. Cohen and J. Young (eds) *The Manufacture of News*, revised edition. London: Constable.
Hall, S., Critcher, C., Jefferson, T., Clarke, J. and Roberts, B. (1978) *Policing the Crisis: Mugging, the State, and Law and Order*. London: Macmillan.
Hannigan, J.A. (1995) *Environmental Sociology: A Social Constructionist Perspective*. London: Routledge.
Hansen, A. (1991) The media and the social construction of the environment, *Media, Culture and Society*, 13: 443–58.
Hansen, A. (ed.) (1993) *The Mass Media and Environmental Issues*. Leicester: Leicester University Press.
Hansen, A. (1994) Journalistic practices and science reporting in the British press, *Public Understanding of Science*, 3: 111–34.
Hansen, A. (2000) Claims-making and framing in British newspaper coverage of the 'Brent Spar' controversy, in S. Allan, B. Adam and C. Carter (eds) *Environmental Risks and the Media*. London and New York: Routledge.
Haraway, D.J. (1991) *Simians, Cyborgs, and Women*. London: Free Association Books.
Hargreaves, I. and Ferguson, G. (2000) *Who's Misunderstanding Whom? Bridging the Gulf of Understanding between the Public, the Media and Science*. London: ESRC / British Academy.
Harré, R., Brockmeier, J. and Mühlháusler, P. (1999) *Greenspeak: A Study of Environmental Discourse*. London: Sage.
Harris, J. (1998) *Clones, Genes and Immortality*. Oxford: Oxford University Press.
Harris, R.F. (1997) Toxics and risk reporting, in D. Blum and M. Knudson (eds) *A Field Guide for Science Writers*. New York: Oxford University Press.
Harrison, T., Projansky, S., Ono, K.A. and Helford, E.R. (eds) (1996) *Enterprise Zones: Critical Positions on Star Trek*. Boulder, CO: Westview.
Hartouni, V. (1997) *Cultural Conceptions: On Reproductive Technologies and the Remaking of Life*. Minneapolis, MN: University of Minnesota Press.
Hartwell, D.G. (1996) *Age of Wonders*. New York: Tom Doherty Associates.
Hawking, S. (1995) Foreword, in L. Krauss, *The Physics of Star Trek*. London: Flamingo.
Hayles, N.K. (1999) *How We Became Posthuman*. Chicago: University of Chicago Press.
Haynes, R.D. (1994) *From Faust to Strangelove: Representations of the Scientist in Western Literature*. Baltimore, MD: Johns Hopkins University Press.
Henriksen, E.K. and Frøyland, M. (2000) The contribution of museums to scientific

literacy: views from audiences and museum professionals, *Public Understanding of Science*, 9: 393–415.
Hirsch, W. (1962) The image of the scientist in science fiction: a content analysis, in B. Barber and W. Hirsch (eds) *The Sociology of Science*. New York: Macmillan.
Hopkin, K. (2001) The risks on the table, *Scientific American*, April.
Hornig, S. (1993) Reading risk: public response to print media accounts of technological risk, *Public Understanding of Science*, 2: 95–109.
Hornig Priest, S. (1999) Popular beliefs, media, and biotechnology, in S. Friedman, S. Dunwoody and C.L. Rogers (eds) *Communicating Uncertainty*. Mahwah, NJ: Lawrence Erlbaum.
Hornig Priest, S. (2001) *A Grain of Truth: The Media, the Public, and Biotechnology*. Lanham, MD: Rowman and Littlefield.
Horton, R. (1999) Genetically modified foods: 'absurd' concern or welcome dialogue?, *Lancet*, 354: 1312.
Howenstine, E. (1987) Environmental reporting: shift from 1970 to 1982, *Journalism Quarterly*, 64(4): 842–6.
Howley, K. (2001) Spooks, spies, and control technologies: technologies in *The X-Files*, *Television and New Media*, 2(3): 257–80.
Hurley, M. (2001) Strategic and conceptual issues for community-based, HIV/AIDS treatments media, Australian Research Centre in Sex, Health and Society.
Huxford, J. (2000) Framing the future: science fiction frames and the press coverage of cloning, *Continuum*, 14(2): 187–99.
Huxley, A. ([1932] 1955) *Brave New World*. Harmondsworth: Penguin.
Irwin, A. (1995) *Citizen Science*. London: Routledge.
Irwin, A. (2001) *Sociology and the Environment*. Cambridge: Polity.
Irwin, A. and Wynne, B. (eds) (1996) *Misunderstanding Science? The Public Reconstruction of Science and Technology*. Cambridge: Cambridge University Press.
Irwin, A., Allan, S. and Welsh, I. (2000) Nuclear risks: three problematics, in B. Adam, U. Beck and J. van Loon (eds) *The Risk Society and Beyond*. London: Sage.
James, E. (1994) *Science Fiction in the 20th Century*. Oxford: Oxford University Press.
Jarmul, D. (1997) Op-ed writing, in D. Blum and M. Knudson (eds) *A Field Guide for Science Writers*. New York: Oxford University Press.
Jasanoff, S. (1997) Civilization and madness: the great BSE scare of 1996, *Public Understanding of Science*, 6: 221–32.
Jasanoff, S. (1998) Book review, *Public Understanding of Science*, 7: 354–6.
Jones, R.A. (1997) The boffin: a stereotype of scientists in post-war British films (1945–1970), *Public Understanding of Science*, 6: 31–48.
Karpf, A. (1988) *Doctoring the Media*. London: Routledge.
Kavanagh, G. (1992) Dreams and nightmares: science museum provision in Britain, in J. Durant (ed.) *Museums and the Public Understanding of Science*. London: Science Museum.
Kiernan, V. (1997) Ingelfinger, embargoes, and other controls on the dissemination of science news, *Science Communication*, 18(4): 297–319.

Kiernan, V. (2000) The mars meteorite: a case study in controls on dissemination of science news, *Public Understanding of Science*, 9: 15–41.

King, G. and Krzywinska, T. (2000) *Science Fiction Cinema*. London: Wallflower.

Kinsella, J. (1989) *Covering the Plague: AIDS and the American Media*. New Brunswick, NJ: Rutgers University Press.

Knight, P. (2000) *Conspiracy Culture: From Kennedy to the X Files*. London: Routledge.

Kolata, G. (1997) *Clone: The Road to Dolly and the Path Ahead*. London: Allen Lane.

Krauss, L. (1995) *The Physics of Star Trek*. London: Flamingo.

Krug, G.J. (1993) The day the earth stood still: media messages and local life in a predicted Arkansas earthquake, *Critical Studies in Mass Communication*, 10: 273–85.

Labinger, J.A. and Collins, H. (eds) (2001) *The One Culture?* Chicago: University of Chicago Press.

Lacey, C. and Longman, D. (1997) *The Press as Public Educator*. Luton: University of Luton Press/John Libbey.

LaFollette, M.C. (1990) *Making Science Our Own: Public Images of Science 1910–1955*. Chicago: University of Chicago Press.

Lambourne, R. (1999) Science fiction and the communication of science, in E. Scanlon, E. Whitelegg and S. Yates (eds) *Communicating Science: Contexts and Channels*. London: Routledge.

Lambourne, R., Shallis, M. and Shortland, M. (1990) *Close Encounters? Science and Science Fiction*. Bristol: Adam Hilger.

Lash, S., Szerszynski, B. and Wynne, B. (eds) (1996) *Risk, Environment and Modernity*. London: Sage.

Lavery, D., Hague, A. and Cartwright, M. (1996a) Introduction, in D. Lavery, A. Hague and M. Cartwright (eds) *Deny All Knowledge: Reading the X-Files*. London: Faber and Faber.

Lavery, D., Hague, A. and Cartwright, M. (eds) (1996b) *Deny All Knowledge: Reading the X-Files*. London: Faber and Faber.

Leiss, W. and Chociolko, C. (1994) *Risk and Responsibility*. Montreal: McGill-Queen's University Press.

Lindahl Elliot, N. (2001) Signs of anthropomorphism: the case of natural history television documentaries, *Social Semiotics*, 11(3): 289–305.

Linné, O. (1991) Journalistic practices and news coverage of environmental issues, *Nordicom Review of Nordic Mass Communication Research*, 1: 1–7.

Long, M., Boiarsky, G. and Thayer, G. (2001) Gender and racial counter-stereotypes in science education television: a content analysis, *Public Understanding of Science*, 10: 255–69.

Long, T.L. (2000) Plague of pariahs: AIDS 'zines, and the rhetoric of transgression, *Journal of Communication Inquiry*, 24(4): 401–11.

Lowe, P. and Morrison, D. (1984) Bad news or good news: environmental politics and the mass media, *Sociological Review*, 32(1): 75–90.

Lowry, B. (1995) *The Truth is Out There: The Official Guide to the X Files*. London: HarperCollins.

Luke, T.W. (1987) Chernobyl: the packaging of transnational ecological disaster, *Critical Studies in Mass Communication*, 4: 351–75.

Lupton, D. (1994) *Moral Threats and Dangerous Desires*. London: Taylor & Francis.

McArthur, L.C. (1998) Report: the portrayal of women in science books for junior readers, *Science Communication*, 20(2): 247–61.

McCurdy, H.E. (1997) *Space and the American Imagination*. Washington, DC: Smithsonian Institution Press.

Macdonald, S. (1996) Authorising science: public understanding of science in museums, in A. Irwin and B. Wynne (eds) *Misunderstanding Science? The Public Reconstruction of Science and Technology*. Cambridge: Cambridge University Press.

Macdonald, S. (1998) Exhibitions of power and powers of exhibition, in S. Macdonald (ed.) *The Politics of Display*. London: Routledge.

McGuigan, J. (1999) *Modernity and Postmodern Culture*. Buckingham: Open University Press.

Macnaghten, P. and Urry, J. (1998) *Contested Natures*. London: Sage.

McNair, B. (1988) *Images of the Enemy*. London: Routledge.

McNeil, M. (1987) *Gender and Expertise*. London: Free Association Books.

McNeil, M. and Franklin, S. (1991) Science and technology: questions for cultural studies and feminism, in S. Franklin, C. Lury and J. Stacey (eds) *Off-Centre: Feminism and Cultural Studies*. London: HarperCollins.

McNeil, M., Varcoe, I. and Yearley, S. (1990) *The New Reproductive Technologies*. Basingstoke: Macmillan.

McRobbie, A. (1994) *Postmodernism and Popular Culture*. London: Routledge.

Maddox, J. (1991) Basketball, AIDS and education, *Nature*, 354(6349): 103.

Major, A.M. and Atwood, L.E. (1997) Changes in media credibility when a predicted disaster doesn't happen, *Journalism and Mass Communication Quarterly*, 74(4): 797–813.

Malone, R.E., Boyd, E. and Bero, L.A. (2000) Science in the news: journalists' constructions of passive smoking as a social problem, *Social Studies of Science*, 30(5): 713–35.

Mazur, A. (1981) Media coverage and public opinion on scientific controversies, *Journal of Communication*, 31(2): 106–15.

Meek, J. (2001) What's eating you?, *The Guardian*, 22 June.

Mellor, F. (2001a) Gender and the communication of physics through multimedia, *Public Understanding of Science*, 10: 271–91.

Mellor, F. (2001b) Between 'fact' and 'fiction': demarcating science from non-science in popular physics books. Unpublished research paper. London: Imperial College of Science, Technology and Medicine.

Mellor, F. (2001c) Colliding worlds: asteroid science and science fiction. Unpublished research presentation, Department of Science and Technology Studies, Cornell University, 5 March.

Miller, D., Kitzinger, J., Williams, K. and Beharrell, P. (1998) *The Circuit of Mass Communication*. London: Sage.

Miller, M.M. and Riechert, B.P. (2000) Interest group strategies and journalistic norms: news media framing of environmental issues, in S. Allan, B. Adam and C. Carter (eds) *Environmental Risks and the Media*. London: Routledge.

Miller, S. (2001) Public understanding of science at the crossroads, *Public Understanding of Science*, 10: 115–20.

Mitchell, W.J.T. (1998) *The Last Dinosaur Book*. Chicago: University of Chicago Press.

Moeller, S.D. (1999) *Compassion Fatigue*. New York: Routledge.

Molotch, H. and Lester, M. (1974) News as purposive behaviour: on the strategic use of routine events, accidents and scandals, *American Sociological Review*, 39(1): 101–12.

Morton, A. (1988) Tomorrow's yesterdays: science museums and the future, in R. Lumley (ed.) *The Museum Time Machine*. London: Routledge.

Mulkay, M. (1997) *The Embryo Research Debate: Science and the Politics of Reproduction*. Cambridge: Cambridge University Press.

Murray, J. (1991) Bad press: representations of AIDS in the media, *Cultural Studies from Birmingham*, 1: 29–51.

Myers, G. (1990) *Writing Biology: Texts in the Social Construction of Scientific Knowledge*. Madison, WI: University of Wisconsin Press.

Nelkin, D. (1995) *Selling Science: How the Press Covers Science and Technology*, 2nd edn. New York: W.H. Freeman.

Nelkin, D. and Lindee, M.S. (1995) *The DNA Mystique*. New York: W.H. Freeman.

Neresini, F. (2000) And man descended from the sheep: the public debate on cloning in the Italian press, *Public Understanding of Science*, 9: 359–82.

Neuzil, M. and Kovarik, W. (1996) *Mass Media and Environmental Conflict: America's Green Crusades*. Thousand Oaks, CA: Sage.

Nieman, A. (2000) The popularisation of physics: boundaries of authority and the visual culture of science. Unpublished PhD thesis, University of the West of England.

North, R.D. (1998) Reporting the environment: single issue groups and the press, *Contemporary Issues in British Journalism*, The 1998 Vauxhall Lectures. Cardiff: Centre for Journalism Studies, Cardiff University.

Nottingham, S. (2000) *Screening DNA: Exploring the Cinema–Genetics Interface*. Stevenage: DNA Books.

Nowotny, H., Scott, P. and Gibbons, M. (2001) *Re-Thinking Science*. Cambridge: Polity.

O'Brien, D. (2000) *SF: UK*. London: Reynolds and Hearn.

Parlour, J.W. and Schatzow, S. (1978) The mass media and public concern for environmental problems in Canada, 1960–1972, *International Journal of Environmental Studies*, 13: 9–17.

Parrinder, P. (1980) *Science Fiction: Its Criticism and Teaching*. London: Methuen.

Parrinder, P. (1990) Scientists in science fiction: enlightenment and after, in R. Garnett and R.J. Ellis (eds) *Science Fiction Roots and Branches*. London: Macmillan.

Patton, C. (1985) *Sex and Germs*. Boston, MA: South End Press.

Pearson, G., Pringle, S.M. and Thomas, J.N. (1997) Scientists and the public understanding of science, *Public Understanding of Science*, 6: 279–89.

Pence, G.E. (1998) *Who's Afraid of Human Cloning?* Lanham, MD: Rowman and Littlefield.

Penley, C. (1997) *NASA / TREK: Popular Science and Sex in America*. London: Verso.

Perlman, D. (1997) Introduction, in D. Blum and M. Knudson (eds) *A Field Guide for Science Writers*. New York: Oxford University Press.

Persson, P. (2000) Science centers are thriving and going strong!, *Public Understanding of Science*, 9: 449–60.

Petit, C. (1997) Covering earth sciences, in D. Blum and M. Knudson (eds) *A Field Guide for Science Writers*. New York: Oxford University Press.

Phillips, Lord (2000) *The Inquiry into BSE and variant CJD in the United Kingdom*. London: HMSO.

POST (Parliamentary Office of Science and Technology) (1995) *Public Attitudes to Science*. Report 69. London: House of Commons. www.parliament.the-stationery-office.co.uk

POST (Parliamentary Office of Science and Technology) (1996) *Safety in Numbers?* Report 81. London: House of Commons. www.parliament.the-stationery-office.co.uk

POST (Parliamentary Office of Science and Technology) (2000a) *The 'Great GM Food Debate'*. Report 138. London: House of Commons. www.parliament.the-stationery-office.co.uk

POST (Parliamentary Office of Science and Technology) (2000b) *Science Centres*. Report 143. London: House of Commons. www.parliament.the-stationery-office.co.uk

Pounds, M.C. (1999) *Race in Space: The Representation of Ethnicity in* Start Trek *and* Star Trek: The Next Generation. Lanham, MD: Scarecrow Press.

Powell, D. (2001) Mad cow disease and the stigmatization of British beef, in J. Flynn, P. Slovic and H. Kunreuther (eds) *Risk, Media and Stigma*. London: Earthscan.

Powell, D. and Leiss, W. (1997) *Mad Cows and Mother's Milk: The Perils of Poor Risk Communication*. Montreal and Kingston: McGill-Queen's University Press.

Priest, S.H. (2001) Cloning: a study in news production, *Public Understanding of Science*, 10: 59–69.

Pyenson, L. and Sheets-Pyenson, S. (1999) *Servants of Nature*. London: Fontana.

Quin, M. (1993) Clone, hybrid or mutant? The evolution of European science museums, in J. Durant and J. Gregory (eds) *Science and Culture in Europe*. London: Science Museum.

Ratzan, S.C. (ed.) (1998) *The Mad Cow Crisis*. London: UCL Press.

Redman, P. (1991) Invasion of the monstrous others: identity, genre and HIV, *Cultural Studies from Birmingham*, 1: 8–28.

Reilly, J. and Miller, D. (1997) Scaremonger or scapegoat? The role of the media in the emergence of food as a social issue, in P. Caplan (ed.) *Food, Health and Identity*. London: Routledge.

Reiss, M.J. and Straughan, R. (1996) *Improving Nature?* Cambridge: Cambridge University Press.

Rensberger, B. (1997) Covering science for newspapers, in D. Blum and M. Knudson (eds) *A Field Guide for Science Writers*. New York: Oxford University Press.

Rhodes, R. (1998) *Deadly Feasts*. London: Touchstone.

Roberts, A. (2000) *Science Fiction*. London: Routledge.

Robertson, G., Mash, M., Tiekner, L., *et al.* (eds) (1996) *FutureNatural: Nature/Science/Culture*. London: Routledge.

Ropeik, D. (1997) Reporting news, in D. Blum and M. Knudson (eds) *A Field Guide for Science Writers*. New York: Oxford University Press.

Rosen, J. (1999) *What Are Journalists For?* New Haven, CT: Yale University Press.

Rosenberg, T. (2001) Look at Brazil, *New York Times*, 28 January.

Rosenthal, E. (2001) China now facing an AIDS epidemic, a top aide admits, *New York Times*, 24 August.

Ross, A. (1991) *Strange Weather*. New York: Verso.

Ross, A. (ed.) (1996) *Science Wars*. Durham, NC: Duke University Press.

Rowe, D. (1997) Apollo undone: the sports scandal, in J. Lull and S. Hinerman (eds) *Media Scandals*. Cambridge: Polity.

Rubin, D.M. (1987) How the news media reported on Three Mile Island and Chernobyl, *Journal of Communication*, 37(3): 42–57.

Rutherford, P. (1999) Ecological modernization and environmental risk, in E. Darier (ed.) *Discourses of the Environment*. Oxford: Blackwell.

Saari, M-A., Gibson, C. and Osler, A. (1998) Endangered species: science writers in the Canadian daily press, *Public Understanding of Science*, 7: 61–81.

Sachsman, D.B. (1976) Public relations influence on coverage of environment in San Francisco area, *Journalism Quarterly*, 53(1): 54–60.

Sack, K. (2001) AIDS epidemic takes toll on black women, *New York Times*, 3 July.

Sagan, C. (1993) Science and pseudo-science, in D. Jarmul (ed.) *Headline News, Science Views II*. Washington, DC: National Academy Press.

Sagan, C. (1997) *The Demon-Haunted World: Science as a Candle in the Dark*. London: Headline.

Salisbury, D.F. (1997) Colleges and universities, in D. Blum and M. Knudson (eds) *A Field Guide for Science Writers*. New York: Oxford University Press.

Salomone, K.L., Greenberg, M.R., Sandman, P.M. and Sachsman, D.B. (1990) A question of quality: how journalists and news sources evaluate coverage of environmental risk, *Journal of Communication*, 40(4): 117–30.

Sammon, P.M. (1996) *Future Noir: The Making of Blade Runner*. London: Orion.

Sandman, P.M. and Paden, M. (1979) At Three Mile Island, *Columbia Journalism Review*, 18(2): 43–58.

Scanlon, E., Whitelegg, E. and Yates, S. (eds) (1999) *Communicating Science, Reader 2*. London: Routledge.

Schoene-Harwood, B. (2000) *Mary Shelley: Frankenstein*. Cambridge: Icon.

Schoenfeld, A.C. (1980) Newspersons and the environment today, *Journalism Quarterly*, 57(3): 456–62.

Schoenfeld, A.C., Meier, R.F. and Griffin, R.J. (1979) Constructing a social problem: the press and the environment, *Social Problems*, 27(1): 38–61.

Select Committee on Science and Technology (House of Lords) (2000) *Science and Society*. London: HMSO.

Shachar, O. (2000) Spotlighting women scientists in the press: tokenism in science journalism, *Public Understanding of Science*, 9: 347–58.

Shamos, M.H. (1995) *The Myth of Scientific Literacy*. New Brunswick, NJ: Rutgers University Press.

Shelley, M. ([1818] 1994) *Frankenstein*. London: Puffin.

Shipman, D. (1985) *Science Fiction Films*. London: Hamlyn.

Shortland, M. and Gregory, J. (1991) *Communicating Science: A Handbook*. London: Longman.

Silver, L.M. (1998) *Remaking Eden: Cloning, Genetic Engineering and the Future of Humankind?* London: Phoenix.

Silverstone, R. (1992) The medium is the museum: on objects and logics in times and spaces, in J. Durant (ed.) *Museums and the Public Understanding of Science*. London: Science Museum.

Simon, A. (1999) *Monsters, Mutants and Missing Links: The Real Science behind The X-Files*. London: Ebury.

Smith, C. (1992) *Media and Apocalypse: News Coverage of the Yellowstone Forest Fires,* Exxon Valdez *Oil Spill, and Loma Prieta Earthquake*. Westport, CT: Greenwood.

Smith, J. (ed.) (2000) *The Daily Globe: Environmental Change, the Public and the Media*. London: Earthscan.

Snow, C.P. (1965) *The Two Cultures: A Second Look*. Cambridge: Cambridge University Press.

Solow, H.F. and Justman, R.H. (1996) *Inside Star Trek*. New York: Pocket Books.

Sontag, S. (1989) *AIDS and its Metaphors*. Harmondsworth: Penguin.

Sood, R., Stockdale, G. and Rogers, E.M. (1987) How the news media operate in natural disasters, *Journal of Communication*, 37(3): 27–41.

Soper, K. (1995) *What is Nature?* Oxford: Blackwell.

Spigel, L. (1997) White flight, in L. Spigel and M. Curtain (eds) *The Revolution Wasn't Televised*. New York: Routledge.

Spinardi, G. (1997) Aldermaston and British nuclear weapons development: testing the 'Zuckerman Thesis', *Social Studies of Science*, 27: 547–82.

Stark, S.D. (1997) *Glued to the Set*. New York: Free Press.

Steinke, J. and Long, M. (1996) A lab of her own? Portrayals of female characters on children's educational science programs, *Science Communication*, 18(2): 91–115.

Stephens, M. and Edison, B.G. (1982) News media coverage of issues during the accident at Three-Mile Island, *Journalism Quarterly*, 59(2): 199–204.

Stocking, H. and Leonard, J.P. (1990) The greening of the media, *Columbia Journalism Review*, December: 37–44.

Sturken, M. (1997) *Tangled Memories*. Berkeley, CA: University of California Press.

Sullivan, R. (1992) Museums, in B.V. Lewenstein (ed.) *When Science Meets the Public*. Washington, DC: AAAS.

Szerszynski, B. and Toogood, M. (2000) Global citizenship, the environment and the media, in S. Allan, B. Adam and C. Carter (eds) *Environmental Risks and the Media*. London and New York: Routledge.

Thompson, K. (1998) *Moral Panics*. London: Routledge.

Toner, M. (1997) Introduction, in D. Blum and M. Knudson (eds) *A Field Guide for Science Writers*. New York: Oxford University Press.

Toumey, C.P. (1996) Conjuring science in the case of cold fusion, *Public Understanding of Science*, 5: 121–33.

Trafford, A. (1997) Critical coverage of public health and government, in D. Blum and M. Knudson (eds) *A Field Guide for Science Writers*. New York: Oxford University Press.

Treichler, P.A. (1999) *How to Have Theory in an Epidemic*. Durham, NC: Duke University Press.

Tulloch, J. and Jenkins, H. (1995) *Science Fiction Audiences*. London: Routledge.

Tulloch, J. and Lupton, D. (1997) *Television, AIDS and Risk*. St Leonards, NSW: Allen and Unwin.

Tulloch, J. and Lupton, D. (2001) Risk, the mass media and personal biography: revisiting Beck's 'knowledge, media and information society', *European Journal of Cultural Studies*, 4(1): 5–27.

Turney, F. (1998) *Frankenstein's Footsteps: Science, Genetics and Popular Culture*. New Haven, CT: Yale University Press.

Ungar, S. (2001) Moral panic versus the risk society: the implications of the changing sites of social anxiety, *British Journal of Sociology*, 52(2): 271–91.

Van Dijck, J. (1995) *Manufacturing Babies and Public Consent*. London: Macmillan.

Van Dijck, J. (1998) *Imagenation: Popular Images of Genetics*. London: Macmillan.

van Loon, J. (2000) Mediating the risks of virtual environments, in S. Allan, B. Adam and C. Carter (eds) *Environmental Risks and the Media*. London and New York: Routledge.

Watanabe, M.E. (2001) AIDS 20 years later. . ., *The Scientist*, 15(12).

Watney, S. (1988) AIDS, 'moral panic' theory and homophobia, in P. Aggleton and H. Homans (eds) *Social Aspects of Aids*. London: Falmer.

Watney, S. (1997) *Policing Desire*, 3rd edn. London: Cassell.

Watney, S. and Gupta, S. (1990) The rhetoric of AIDS, in T. Boffin and S. Gupta (eds) *Ecstatic Antibodies*. London: Rivers Oram Press.

Watson, J.D. (2000) *A Passion for DNA*. Oxford: Oxford University Press.

Watson, M.A. (1990) *The Expanding Vista: American Television in the Kennedy Years*. Durham, NC: Duke University Press.

Weeks, J. (1985) *Sexuality and its Discontents*. London: Routledge and Kegan Paul.

Weeks, J. (1993) AIDS and the regulation of sexuality, in V. Berridge and P. Strong (eds) *AIDS and Contemporary History*. Cambridge: Cambridge University Press.

Wellings, K. (1988) Perceptions of risk: media treatment of AIDS, in P. Aggleton and H. Homans (eds) *Social Aspects of AIDS*. London: Falmer.

Wells, H.G. ([1898] 1993) *The War of the Worlds*. London: Everyman.
Wells, H.G. ([1901] 2001) *The First Men in the Moon*. London: Gollancz.
Welsh, I. (2000) *Mobilising Modernity: The Nuclear Moment*. London: Routledge.
Whitfield, S.E. and Roddenberry, G. (1968) *The Making of Star Trek*. New York: Ballantine.
Wiegman, O., Gutteling, J.M., Boer, H. and Houwen, R.J. (1989) Newspaper coverage of hazards and the reactions of readers, *Journalism Quarterly*, 66(4): 846–52.
Wilkins, L. and Patterson, P. (1987) Risk analysis and the construction of news, *Journal of Communication*, 37(3): 80–92.
Wilkins, L. and Patterson, P. (1990) Risky business: covering slow-onset hazards as rapidly developing news, *Political Communication and Persuasion*, 7(1): 11–23.
Williams, R. ([1958] 1989) Culture is ordinary, in R. Gable (ed.) *Resources of Hope*. London: Verso.
Williams, R. (1983) *Keywords*. London: Flamingo.
Wilson, A. (1992) *The Culture of Nature: North American Landscape from Disney to the Exxon Valdez*. Cambridge, MA: Blackwell.
Wilson, C. (2000) Communicating climate change through the media: predictions, politics and perceptions of risk, in S. Allan, B. Adam and C. Carter (eds) *Environmental Risks and the Media*. London and New York: Routledge.
Winter, G. (2001) Contaminated food makes millions ill despite advances, *New York Times*, 18 March.
Wolmark, J. (ed.) (1999) *Cybersexualities*. Edinburgh: Edinburgh University Press.
Wykes, M. (2000) The burrowers: news about bodies, tunnels and green guerillas, in S. Allan, B. Adam and C. Carter (eds) *Environmental Risks and the Media*. London and New York: Routledge.
Wynne, B. (1992) Public understanding of science research: new horizons or hall of mirrors?, *Public Understanding of Science*, 1: 37–43.
Yeo, R. (1993) *Defining Science: William Whewell, Natural Knowledge and Public Debate in Early Victorian Britain*. Cambridge: Cambridge University Press.
Young, P. (1997) Writing articles from science journals, in D. Blum and M. Knudson (eds) *A Field Guide for Science Writers*. New York: Oxford University Press.
Ziman, J. (1992) Not knowing, needing to know, and wanting to know, in B.V. Lewenstein (ed.) *When Science Meets the Public*. Washington, DC: AAAS.
Ziman, J. (2000) *Real Science*. Cambridge: Cambridge University Press.

INDEX

GENETICS AND SOCIETY

Alison Pilnick

> With panoramic coverage and accessible style, one cannot help but recommend this volume to students, lecturers and researchers. It will prove an indispensable textbook and source of reference. From science to ethics, medicine to agriculture, and disability to cloning, Alison Pilnick provides a highly readable and truly informative account of the impact of modern genetics on contemporary social life.
>
> Lindsay Prior, Reader in Sociology, Cardiff University, and Director of the Health and Risk Programme at the University of Wales College of Medicine

- What impact do advances in genetic science have on the relationships between humans and the world in which they live?
- How does social context affect the development and implementation of genetic research?
- How can the social sciences be used to develop a critical perspective on advances in genetics?

This is a book about contemporary developments in the scientific understanding of genetics, and the ways in which these are transforming possible relations between humans and the world around them. It is the first book of its kind, aiming to encourage readers to critically examine social issues that relate to genetic science and practice of genetic science. The focus is mainly, though not exclusively, on human genetics, exploring those developments which are seen as most significant in terms of public perceptions, social impact, or public policy. It covers a wide range of current and potential applications of genetic science and is clearly and accessibly written, assuming no prior knowledge on the part of the reader. Instead, genetic knowledge is placed in its social context.

Contents

224pp 0 335 20735 9 (Paperback) 0 335 20736 7 (Hardback)

CULTURES OF POPULAR MUSIC

Andy Bennett

- What is the relationship between youth culture and popular music?
- How have they evolved since the second world war?
- What can we learn from a global perspective?

In this lively and accessible text, Andy Bennett presents a comprehensive cultural, social and historical overview of post-war popular music genres, from rock 'n' roll and psychedelic pop, through punk and heavy metal, to rap, rave and techno. Providing a chapter by chapter account, Bennett also examines the style-based youth cultures to which such genres have given rise. Drawing on key research in sociology, media studies and cultural studies, the book considers the cultural significance of respective post-war popular music genres for young audiences, with reference to issues such as space and place, ethnicity, gender, creativity, education and leisure. A key feature of the book is its departure from conventional Anglo-American perspectives. In addition to British and US examples, the book refers to studies conducted in Germany, Holland, Sweden, Israel, Australia, New Zealand, Mexico, Japan, Russia and Hungary, presenting the cultural relationship between youth culture and popular music as a truly global phenomenon.

Contents

Introduction – Post-war youth and rock 'n' roll – Sixties rock, politics and the counter-culture – Heavy metal – Punk and punk rock – Reggae and Rasta culture – Rap music and hip hop culture – Bhangra and contemporary Asian dance music – Contemporary dance music and club culture – Youth and music-making – Whose generation? Youth, music and nostalgia – Glossary – References – Index.

208pp 0 335 20250 0 (Paperback) 0 335 20251 9 (Hardback)

MASCULINITIES AND CULTURE

John Beynon

- What is 'masculinity'? Is 'masculinities' a more appropriate term?
- How are masculinities socially, culturally and historically shaped?
- How are particular masculinities created, enacted and represented in specific settings?
- How can masculinities best be researched and theorized?

Masculinities and Culture explores how 'masculinities', or ways of 'being a man', are anchored in time and place; the products of socio-historical and cultural circumstances. It examines the emergence of a masculinity fit for Empire in the mid to late nineteenth century and, by way of contrast, the more recent media-driven, commercial New Man and New Lad masculinity. The author considers some of the media discourses shaping masculinities today, and the formation of specific masculinities in specific settings (such as prisons, hospitals and schools) which both define, and in turn are defined by, strongly held conceptions of acceptable masculine behaviour. He concludes by reviewing a range of ways in which masculinities might be researched, from fieldwork and auto/biographical and life history approaches through to semiotics and the use of both film and literary texts. This lively text provides a comprehensive introduction to contemporary debates concerning masculinities as gendered constructions, along with the means of researching and theorizing them.

Contents

208pp 0 335 19988 7 (Paperback) 0 335 19989 5 (Hardback)

CINEMA AND CULTURAL MODERNITY

Gill Branston

- What is the relationship of popular cinema to the concept of 'modernity'?
- What now are the key areas of debate which focus the study of cinema and its audiences?
- How can we understand the relationship of cinema to both the pleasures of consumerism and the inequalities addressed by critical politics?

Cinema and Cultural Modernity carves a lucid path through the central debates of film and cinema studies and explores these in their social and political contexts. The book includes histories of the ways in which we view Hollywood's global dominance, up to the development of late modernity and the declaration of 'postmodernity'. In an accessible fashion, it discusses changing theorizations of the economics, audiences, and fascinations of cinema, addressing concepts such as agency, negotiation and identification, and global 'popularity' within contemporary cultures of celebrity, consumption and the visual. Gill Branston outlines the need for cinema study that is both sensitive to the formal 'textiness' of films, but also less anxious about arguing for its position within broad agendas of representation. At the same time, the author links such areas to both the pleasures of consumption, which cinema so often evokes and embodies, and to the need for a new, critical politics to address the persistent inequalities of modernity, inequalitiies which still fuel lively interest in questions of representation. The result is an incisive text for undergraduate courses and an essential reference for researchers.

Contents

224pp 0 335 20076 1 (Paperback) 0 335 20077 X (Hardback)

NEWS CULTURE

Stuart Allen

- In what ways do the news media reproduce the social divisions and hierarchies of modern societies?
- Can jounalists be 'objective' in their reporting? How did the conventions of 'objective' reporting become established in the first place?
- How do people make sense of the news in relation to their everyday life? Is journalism a form of popular culture?

News Culture provides a rich and lively discussion, full of insights into the changing forms, practices, institutions and audiences of journalism. Its fresh engagement with a wide-ranging number of issues, together with the use of thought-provoking examples, offers the reader a comprehensive assessment of different critical approaches to the news media on both sides of the Atlantic.

The book begins with an historical consideration of the rise of 'objective' reporting in newspaper, radio and televisual journalism. It goes on to explore the way news is produced, its textual conventions as a genre of discourse, and its negotiation by the reader, listener or viewer as part of everyday life. Attention then turns to address the cultural dynamics of sexism and racism as they shape different instances of news coverage. Finally, the book examines ongoing debates about the status of journalism as a form of popular culture.

News Culture will be welcomed as essential reading by students and researchers in cultural studies, media studies, journalism, sociology, politics and criminology.

Contents

240pp 0 335 19956 9 (Paperback) 0 335 19957 7 (Hardback)

openup
ideas and understanding
in social science

www.**openup**.co.uk

Browse, search and order online

Download detailed title information and sample chapters*

*for selected titles

www.**openup**.co.uk